AF744231

UNIVERSITY OF BRADFORD
MAKING KNOWLEDGE WORK™

LIBRARY

This book should be returned not later than the last date stamped below.
The loan may be extended on request provided there is no waiting list.
FINES ARE CHARGED ON OVERDUE BOOKS

ONE WEEK ONLY

L45

°350EE

BD 034423664 0

The Family in the Mediterranean Welfare States

The Family in the Mediterranean Welfare States

MANUELA NALDINI

FRANK CASS
LONDON • PORTLAND, OR

First published in 2003 in Great Britain by
FRANK CASS PUBLISHERS
Crown House, 47 Chase Side
London N14 5BP

and in the United States of America by
FRANK CASS PUBLISHERS
c/o ISBS, 920 NE 58th Avenue, #300
Portland, Oregon 97213-786

Website: www.frankcass.com

Copyright © 2003 Manuela Naldini

British Library Cataloguing in Publication Data

Naldini, Manuela
The family in the Mediterranean welfare states
1. Welfare state – Italy. 2. Welfare state – Spain.
3. Family – Italy. 4. Family – Spain. 5. Italy – Social conditions
– 1994– 6. Spain – Social conditions – 1975–
I. Title
361.9′45
ISBN 0 7146 5230 X (cloth)

Library of Congress Cataloging-in-Publication Data

Naldini, Manuela
The family in the Mediterranean welfare states / Manuela Naldini
p. cm.
Includes bibliographical references and index.
ISBN 0-7146-5230-X (cloth)
1. Family – Italy. 2. Family – Spain. 3. Welfare state –
Italy. 4. Welfare state – Spain. 5. Family policy – Italy.
6. Family policy – Spain. I. Title

HQ630.N35 2003
306.85′0945 – dc21

2002041566

All rights reserved. No part of this publication may be reproduced, stored in or introduced into a retrieval system or transmitted in any form or by any means, electronic, mechanical, photocopying, recording or otherwise, without the prior written permission of the publisher of this book.

Typeset in 10.25/12pt Times New Roman by Frank Cass Publishers
Printed in Great Britain by MPG Books Ltd, Bodmin, Cornwall

Contents

Figures and Tables

Figures

Tables

Foreword

Family arrangements, in terms of gender and intergenerational relations and patterns of obligations, have been at the core of most social policy since the beginning of the modern welfare state. In regulating labour relations and conditions, and in defining which needs might be socially acknowledged and (at least partly) supported, in fact, social legislation and then social policies have implicitly regulated, or at least interfered, with family and household formation models: redefining the relationships of dependence and interdependence not only between genders, but also between generations; modifying the conditions and costs of reproduction; rewarding, or vice versa discouraging, particular patterns of family and kin obligations. An example of this is the introduction of old age pensions at the beginning of the century: having a pension, in fact, allowed the elderly not only to look with a degree of security to their future out of work; it also allowed them not to depend too exclusively on their kin's, particularly their children's, solidarity. On the contrary, restrictions on child and women's labour, together with the introduction of compulsory schooling, constituted *de facto* a means of regulating workers' households, with regard to gender and intergenerational relations: first of all by distinguishing household members between 'workers' and 'family dependants'.

Out of this process, which was by no means linear and homogeneous across countries and across social groups, policies developed some ideal household model premised on the one hand on the presence of a male breadwinner, who was not only responsible for providing income, but was also the mediator of social protection for women and children; on the other hand, on the presence of a wife mother who was responsible both for occasionally integrating needed income, and for systematically providing care, particularly for young children.

The attention to the different ways in which gender relations and intergenerational obligations are supported, expected or promoted by social policies has contributed not only to enlarging the scope and dimensions of welfare state research, but to the creation of a whole new 'family' of welfare state typologies. Thus, on the one hand we have (different) typologies proposed by scholars mainly interested in how gender relations are framed within social policies and welfare mixes (e.g. Lewis, 1997; Sainsbury, 1996; Hobson, 1990; O'Connor *et al.*, 1999; Daly, 2000); on the other hand we have typologies proposed by scholars who focus on how intergenerational relations and obligations are framed by and within social policies (e.g.

Millar and Warman, 1996; Gauthier, 1996; Hantrais and Letablier, 1996).

It should be pointed out that the attention to the way intergenerational obligations are framed both practically and symbolically in welfare state policies is greatly indebted to feminist analysis, insofar as the latter is responsible for having introduced and developed the dimension of caring as an integral part both of welfare state analysis, in terms of needs, obligations, as well as rights (e.g. Knijn and Kramer, 1997; Leira, 1998; Lewis, 1998). Yet these two focuses still remain only partly integrated at the theoretical level, limiting the explanatory power of family arrangements for understanding welfare arrangements. The integration of these two perspectives is instead at the core of Naldini's approach, allowing her to grasp the specificity of the role allocated to family arrangements and solidarity in Mediterranean welfare states. Certainly her national cases could not be understood without focusing on both gender and intergenerational assumptions and arrangements. Yet, one might wonder whether new insights would be gained also with regard to other, better-known welfare types, if her research questions and interpretive lens were applied to them.

Generally, introducing a focus on family and kin relations – from a gender, but also from an intergenerational perspective – has not only enriched but also complicated social policy analysis and comparison, reducing the explanatory power of, albeit useful, typologies. If these vary as soon as we introduce a new dimension/indicator, or change one, one either must rethink the whole set of dimensions and indicators deemed crucial, or question the degree of generalization of typologies themselves. Since this is a relatively new field and yet to be fully explored, maybe one should be satisfied with partial exercises in generalization and in typology construction, without aiming – yet at least – at grand theories and fully coherent and encompassing typologies.

From this perspective, comparative analyses such as that performed by Naldini in this book appear of crucial value: insofar as it focuses on one particular type of welfare state, the so-called Mediterranean one, in order to understand its specific pattern of development and its cultural and institutional assumptions – with specific regard to family arrangements. Analogously to what O'Connor, Orloff and Shaver (1999) have done with regard to 'liberal' welfare states, and with the added value of focusing on the lesser-known and still under-studied ideal type in comparative welfare research, Naldini is not first and foremost interested in contrasting her cases with other more different ones. Rather, she is interested in understanding how they develop 'from within', how they have defined entitlements, rights and obligations over time, how and where they set, and moved, boundaries between individual rights and family membership, family obligations and societal obligations – to some degree how they define 'society' itself *vis-à-vis* both the individual and the family. Thus, her comparative exercise arises from carefully devised national case studies, which link institutional and cultural analysis with regard to gender as well as intergenerational patterns of relationships and obligations.

The end product is not only a magnificent contribution to the study of Mediterranean welfare states, but, first of all methodologically and conceptually, to comparative welfare state research as such.

Chiara Saraceno
University of Turin

Preface

This book analyses from both a historical and a comparative perspective the relationship between the family and the welfare state in two Mediterranean countries: Italy and Spain. Two aims form the focus of the book. First, it attempts to open the 'black box' of the family in welfare state analysis, introducing a focus not only on gender, but also on intergenerational and kin relationships. Second, it aims to explain the paradox of two 'familialist' countries, Italy and Spain, which do not have significant social policies aimed at families with children. To this end I take into account several factors and their complex cultural and political intricacies: the legacy of fascism, the role of the Catholic Church, and the specific role played by leftist parties in defining family policy as labour policy.

The central proposition of this book is that in these two Mediterranean countries the assumptions underlying social policies and legal norms were only to a lesser degree based on the male breadwinner family model, which prevailed in most other European countries. In these two countries certain socio-economic, ideological and cultural features hampered the possibility of completely achieving the male breadwinner family model. Instead, in Italy and Spain the principle of the male breadwinner family model was 'stretched' to include family/kinship solidarity. This model, based on a set of assumptions about the existence of intergenerational ties and kin solidarity throughout the life course and between households, reveals one of the most unique mechanisms by which Mediterranean welfare states function.

Acknowledgements

This book originated when I was a Ph.D. student at the European University Institute at Florence. My dissertation and thus this book would not have been written without the support of a number of people. First of all, I am deeply indebted to Chiara Saraceno, who has stimulated and encouraged me to take on many of the ideas found in this book; her sharp criticism and understanding advice have been crucial contributions to the analytical coherence of this work. She strongly encouraged me to find an editor willing to publish this work. I am indebted to Stefano Bartolini, for his unwavering support during my years at the EUI. His invaluable criticism and advice have contributed to the development of the thesis. I would like to thank Peter Flora, who helped me elaborate the project in its earliest stages. During my tenure at the Institute I had the opportunity to attend interesting seminars and discuss my work with many people. Among them I would like to thank Colin Crouch for having given me the opportunity to discuss my thesis during his fruitful seminars, and later on for having advised me to find an editor for this work.

Many thanks go to Maurizio Ferrera for his helpful comments. I am also indebted to Luis Moreno, who offered thoughtful advice during the drafting of the thesis. I would like to express my gratitude to Lluis Flaquer and Anna Cabré, who helped me during my stay in Barcelona. My thanks go also to my colleagues and friends, particularly to Teresa Jurado, who discussed many theoretical and practical doubts with me, and to Maria José Gonzalez for her advice and support during my recurrent journey in Barcelona. I thank all my friends and flatmates of my last years of research in Florence, Barcelona and Paris: Teresa, Maria José, Laura, Juan, Marta, Gaia, Cristina, Giovanni, Federica, Jean-Pierre, Nina, Sandra, Maria, Bente and Maribel. Many thanks also to the members of the European University Institute, especially to the members of the SPS Department, the staff and English teachers at the Language Centre, and to Jackie Gordon and Sarah Grattan for support in revising the English. During the last months of my work at the University of Turin I have to thank especially Chiara Bertone, who has read the first draft of this book and advised me. Finally, my thanks go to my parents and to Nevio, who have patiently supported me in life and during my long academic journey.

Abbreviations & Acronyms

AES	Acuerdo Economico y Social
ANE	Acuerdo Nacional Economico
AP	Alianza Popular
APC	Atti Parlamentari Camera
APS	Atti Parlamentari Senato
BOCG	Boletin Oficial de las Cortes General
CCOO	Comisiones Obreras
CD	Coalición Democratica
CDi	Congreso de los Diputados
CEOE	Confederación Española de Organizaciones Empresariales
CGIL	Camera Generale Italiana Lavoro
CIF	Comitato Italiano Femminile
CISL	Confederazione Italiana Sindacato Lavoratori
CNAF	Caisse Nationale des Allocations Familiales
DC	Democrazia Cristiana
IM	Instituto de la Mujer
INAM	Istituto Nazionale Assicurazioni sulle Malattie
INP	Instituto Nacional de Previsión
INPS	Instituto Nazionale Previdenza Sociale
IRPEF	Imposta sui Redditi delle Persone Fisiche
IRPF	Impuesto sobre la Renta de las Personas Fisicas
MSI	Movimento Sociale Italiano
ONMI	Opera Nazionale Maternità e Infanzia
OPIM	Obras de Protección a la Infancia y la Maternidad
PCE	Partido Comunista de España
PCI	Partito Comunista Italiano
PDS	Partito Democratico di Sinistra
PLI	Partito Liberale Italiano
PNV	Partido Nacionalista Vasco
PP	Partido Popular
PPI	Partito Popolare Italiano
PRI	Partito Repubblicano Italiano
PSDI	Partito Socialdemocratico Italiano
PSI	Partito Socialista Italiano
PSOE	Partido Socialista Obrero Español
S	Senado
SMI	Salario Minimo Interprofessionale
UCD	Unión de Centro Democrático
UDI	Unione Donne Italiane
UGT	Unión General de Trabajadores
UIL	Unione Italiana Lavoratori

INTRODUCTION

Throughout the Western industrialized world the crisis of the welfare state has raised a new interest in studying the link between the welfare state and the organization of the family. This interest has resulted in a 'rediscovery' of the family. It is not simply an intellectual curiosity or a development of welfare state theory. Rather, the emergence of the 'family question' in welfare state research and thinking parallels policy concerns with regard to how family changes, particularly with regard to gender arrangements and intergenerational patterns of solidarity, may affect existing welfare state arrangements and/or encourage, or hinder welfare state restructuring. Thus, on the one hand, the decline of the male breadwinner norm, the fragmentation of traditional family obligations, and increasing marital instability, together with the ageing of the population and fertility decline, are perceived as one of the causes of the welfare state crisis. On the other hand, the growing awareness of the limits of the welfare state and stronger pro-market ideologies have led to a reappraisal of the role of the market, the family and of voluntary organizations as alternative mechanisms of resource allocation for the provision of social protection and welfare. Now more than ever before the debates about the future of the welfare state are intimately connected with debates about the family. We are, in brief, assisting in the rediscovery of the family, in an era of welfare state crisis. The family has undergone rapid changes, and this has become a problem because, more than ever before, the well-being of individuals and the future of the welfare state depend on what can and should be expected of the family in supporting those in need of financial support and care.

However, the family has not played the same role in protecting those in need in all welfare states, nor are the same family changes undermining the welfare state in all countries. Furthermore, the requirement of and expectation of family obligations vary from country to country. Taking Europe alone, we find that countries such as, Greece, Italy, Spain and Portugal, the so-called, Mediterranean or 'southern' countries in welfare state literature, are often regarded as the more family-oriented or 'familialistic' countries, compared to others where family ties are less important. Clearly, it is assumed that in these countries a fundamental role is reserved for the family and for kin ties, in the functioning of the welfare mix. Indeed, from a comparative perspective intergenerational ties and kin solidarity are strong in these countries. Moreover, family changes, increasing marital instability and the 'de-institutionalization' of marriage are phenomena less evident in southern countries, where demographic changes (fertility decline) are seen as more dramatic and problematic for the functioning of the welfare state.

In looking at what the state does for families, or simply at a state's effort in relation to families, one discovers that the most 'familialist' countries paradoxically are those in which active family policies are extraordinarily underdeveloped. At the same time, in these same countries, family obligations, even those stipulated in law, are more extended than in less 'familialist' ones. Thus it seems that Mediterranean welfare states, which have relied more heavily on traditional extended family obligations, may more easily turn to them in a period of fiscal crisis. Yet they are less well equipped to deal with demographic and behavioural changes in family arrangements.

What, then, makes an analysis of the family and the welfare state in Mediterranean countries particularly interesting? Surely there is nothing new in arguing that the family has played a more important role for the functioning of the welfare state in Italy, Spain, Greece and Portugal than it has in other countries? In this part of Europe, the term *family* in itself has a broad meaning which encompasses more than the nuclear family (the one characterized by the presence of only two generations, that is, a couple cohabiting with children) or the household. Within Mediterranean countries and languages, the term 'family' indicates a diffuse network of relationships, obligations and loyalties. More specifically, boundaries between households and kin networks are often thought of not only in emotional and affective terms, but also in legal and practical ones.

Recent welfare state literature has advanced the idea that these countries (Italy, Greece, Portugal and Spain, having some doubts whether France can also be included) constitute a distinctive and unique cluster for analysis. Most of the scholars have paid attention to the political institutional characteristics of the Mediterranean welfare state model. Some have concentrated on the role of the 'family' as a welfare agency, while others have paid attention to the value system that characterizes the Mediterranean welfare state model. But these analyses have been mostly gender-blind and unfocused when it comes to what is really meant by family.

Less attention has been paid to the ways in which gender roles, intergenerational ties and kin solidarity are embedded in social policies and laws. In contrast, in this book it is argued that attention should be focused not on the family generically, nor only on its gender dimension, but on the way in which gender, intergenerational and kin dimensions have been framed and constructed in social policies and laws. This might help to explain why countries categorized as belonging to the same 'family' of nations, or clustered together in terms of standard welfare state characteristics (pension and unemployment systems), developed along different lines with respect to gender, intergenerational and kin dimensions of welfare state patterns of redistribution.

While it is clear that there are common features which have marked the evolution of most European welfare states, there must be certain qualitative and historical differences which account for the large variations in social policies involving family arrangements, among countries and over time. It is not only the role played by the family in the economy of the welfare state as a whole that can help shed light on these differences. Rather, differences in the way the family – given its gendered and generational lines of division – is conceptualized, acknowledged and strengthened by policies and laws might also be a crucial feature of a specific national welfare regime, and/or of a specific model of the welfare state.

Comparative welfare state studies are greatly indebted to research undertaken from a gender perspective, in that it exposed the implicit and explicit assumptions of different welfare state patterns with regard to gender and family arrangements: model of motherhood, of marriage, of parental obligations, of children's rights and so forth. But most gender studies of social policies have failed to include in their analyses patterns of solidarity and obligations between and within generations, within the household, and amongst kin. The political institutionalization of these norms through law and social policies codifies social reproduction, and relates how a particular society has solved the problem of the relationship between paid and unpaid work.

This book analyses in historical and comparative perspective the relationship between the family and the welfare state in two Mediterranean countries: Italy and Spain. These exhibit many similarities in their political, social and economic development: both experienced fascism, both have been characterized by a strong influence of the Catholic Church, and both demonstrate similar family patterns and similar paths of family change. In addition, and this is what makes an analysis of the welfare state and the family in these two countries particularly interesting, these two Catholic countries, which both in the literature and in general knowledge are considered strongly 'familialistic' paradoxically show the lowest level of public support for families with children in Europe.

The Aims of this Book

This book has two goals. First, it aims at opening the 'black box' of the 'family' in welfare state analysis, introducing a focus, not only on gender, but also on intergenerational and kin relationships and on explicit and implicit obligations. Second, it attempts to explain the paradox of two 'familialist' countries, Italy and Spain, without significant social policies for families with children. The twin interaction between family arrangements and welfare state arrangements will be analysed, focusing on the diversity in the way laws and social policies have defined the family. In particular, this work reconstructs how public and private obligations and responsibilities have been legally defined with regard to the needs of financially *dependent* individuals, on the basis of age, generation, gender and kinship. To this end it is argued that it is necessary to focus on the issue of familial responsibilities and the corresponding questions of 'familial dependence' and of dependent family relationships. In general terms, the way in which the family is defined in social policies and laws is the result of a set of intertwined factors: assumptions about existing family patterns and gender and intergenerational arrangements; the most desirable family model; how families function in society; how the labour market works; and the most desirable boundary between public and private forms of responsibilities. Importantly, these factors change over time.

Sets of interlinked questions are raised in this book. The first set of substantive questions concerns the similarities between Italian and Spanish historical cases regarding social policies related to the family, and their diversity from other European countries. How have Italian and Spanish social policies and laws defined the family? Which types of family models have been implicitly supported by social policy? How have these two countries allocated the tasks between state and family, and within the family, between genders, generations and kin relations? How have they defined dependent family relationships? Finally, to what extent have family ties and kin networks constituted a unique feature of the Mediterranean welfare state model?

Historical and comparative gender studies of the welfare state have shown that the majority of European welfare states have supported and incorporated in to their institutions ideas and expectations about the existence of a male breadwinner family model. According to this model the husband/father is the breadwinner while the woman is devoted to homemaking and caring work. These assumptions have then been incorporated into certain social policies and legal norms which support the male breadwinner family model through income benefits or through the introduction of 'socially derived rights' for women.

One question in this study is to what degree the Italian and Spanish welfare states, which certainly are based on assumptions concerning the gender division of labour in the family, may be defined as male breadwinner models, given their equally strong assumptions concerning extended family obligations.

Behind these questions lurks my key hypothesis, namely that assumptions and principles underlying the definition of the family in Italy and Spain have been based only to a low degree on the pure, nuclear, male breadwinner family model. Rather, my central argument is that in Italian and Spanish historical cases, certain principles of the male breadwinner family model have assumed a different shape and have been 'stretched' to support what I call the 'family/kinship solidarity' model. The core of this model has been constituted, on the one hand, by permanent legal definitions of kin obligations which go beyond the residential boundary of the household to include adult parents, such as sons- and daughters-in-law, fathers- and mothers-in-law and siblings. Strong and extended family obligations are noticeably accompanied by the unchanging idea that caring work is a family responsibility. On the other hand, this model has been characterized by the funnelling of a part of public resources to the male breadwinner, not only for his wife and children, but also for other economically dependent family/kin members. This idea of an extended number of family members to be protected may lead one to conceive of the Mediterranean welfare state as being more 'universalist' in terms of coverage, in a certain sense, than the occupationally based system of the German type. The Italian and Spanish welfare state systems, in fact, stretch the male breadwinner family model in order to include all those family members who are 'outsiders' in relation to the main social protection system, not only spouse and children but also other dependent relatives. These assumptions and this extension of protection from the male breadwinner to the wider family network have surely contributed to reinforcing intergenerational ties and kin solidarity, while weakening individual entitlements.

Another central argument of this book is that, while passing from the authoritarian to the democratic period, the family/kinship solidarity model has been characterized in both countries by a very low level of public support for families with children. Thus, the second aim of this book is to explain why the Italian and Spanish welfare states have historically shown one of the lowest levels of public support for families with children during the democratic period.

In searching for an explanation for the two historical cases, a second set of questions is raised. Has the specific path of modernization undertaken by the southern European countries played a unique role in the definition of its family model? And what have been the roles played by strong family ties and by a strong 'culture' of the family in shaping the type of social policies related to the family?

While specific labour market characteristics, unique processes of modernization, and the legal permanence of intergenerational ties and kin solidarity have been important influences, they do not provide a full explanation for the permanence of the family/kinship solidarity model and a low level of public support to families.

In particular, in the two historical cases further explanatory variables will be put forward, looking at the political and ideological configurations of the two countries. How has the legacy of fascism influenced subsequent social policies related to families? To what extent has the strong presence of the Catholic Church interfered with legislative changes and shaped the allocation of responsibilities between the state and the family in the area of reproduction and caring work? Why did the prevalence of a Catholic culture and of Catholic-inspired political parties permit strong economic support for families with children in countries such as France and Belgium, but not in Italy and Spain? How much can the specific traits of the polities in southern countries, which explain the 'clientelistic–particularistic' characteristics of the welfare states, also explain the low level of public support for families with children?

In the final part of this book I try to demonstrate that the permanence of the family/kinship solidarity model, and in particular the low level of public support for families with children well into the democratic period, can be explained only if we look at the political, ideological and cultural configurations of the two countries. In particular, I will show that the interplay between conflicts over family issues in the political arena, the legacy of fascism, and the strong influence exercised over family matters by the Catholic Church become fundamental factors in explaining both the permanence of the 'family/kinship solidarity' model and low public support to families with children.

PART ONE

ANALYSING FAMILIES AND THE WELFARE STATE

Chapter 1

WELFARE STATE AND FAMILY MODELS

Theoretical Orientations

The theoretical core of this book draws primarily on the literature on comparative welfare states and gender studies. This chapter is divided into four main sections. The first section provides some initial definitions and investigates the relationship between the family and the state. The second provides a historical analysis of when and why research on the sociology of the family and the welfare state converged. In the third section I explore the main traditions of comparative welfare state research, focusing particularly on two new strands of research: the gendered welfare state and the Mediterranean welfare state model. Finally, the fourth section illustrates the theoretical perspective in which this book is placed.

Problems of Definition

Some key concepts in this book are 'family', 'state', 'social policy' and 'welfare state'. The first two are particularly problematic. 'Family' and 'state' cannot be seen as unitary concepts with a single meaning or reference point. There have been different family forms in the past and in modern society; moreover, the concept of 'family' overlaps with concepts such as household, kinship, marriage and parenthood. The concept of family itself has different meanings in different countries, and in some countries may have a broader meaning which is not clearly delimited in the language or daily life of the people. In Italy and Spain, for instance, the single term *la famiglia/la familia* can refer both to individual households and to kinship networks. Different values and ideas also exist about what families should be like, and what kind of living arrangements are acceptable and merit the term 'family'. The phrase 'the family' is polysemic and thus avoided in this work, while alternative terms such as family, families and/or family/kinship networks are used (Finch, 1989). 'State' is also a concept that incorporates many different meanings. The state can be broadly defined as a set of institutions and/or organizations (i.e. central and local governments, courts, police and so on).

Neither does social policy have a single meaning, and perhaps given that this term refers to a great number of public policies that vary across countries and over time, it is more appropriate to use the term 'social policies'. Social policies can be seen as an assemblage of public interventions that have social purposes and effects, including the distribution of life chances, well-being and the quality of life, as well as social consequences of other policies. This definition of social policies clearly inspires Marshall's definition, where

> Social policy is not a technical term with an exact meaning ... it is taken to refer to the policy of governments with regard to the action having a direct impact on the welfare of citizens, by providing them with service and income. The central core consists of social insurance, the health and welfare services, and housing policy. (quoted in Titmuss, 1974: 30)

From this perspective the meaning of the term 'social policy' does not differ greatly from that of 'welfare state'. According to Flora and Heidenheimer, 'welfare and state are among the most ambiguously employed terms in contemporary English political vocabulary' (1981a: 5). Even today, the concept underlying the term 'welfare state' has hardly been defined once and for all (Ferrera, 1996). It is difficult to define because the concept is related to modernization processes and to basic development problems, and it should be interpreted as a response to the demand for socio-economic equality and security. From this point of view

> There are three basic means by which the welfare state pursues its goals: the direct payment of cash benefits, the direct provision of services in kind, and the indirect extension of benefits through tax deductions and credits … The objectives and instruments … do not define the historical core of the welfare state. It has become usual to identify the beginning of the modern welfare state with the innovation of social insurance. (Flora and Heidenheimer, 1981b: 25–7)

From another perspective, the welfare state has been conceptualized as a state committed to modifying the play of social or market forces in order to achieve greater equality (Ruggie, 1984). Recently, Ferrera has tried to summarize the existing suggestions into a new definition of the term welfare state. She argues that the welfare state can be defined as 'an aggregation of public interventions related to modernisation processes, which provides protection under the form of assistance, assurance and social security, introducing specific social rights in the case of pre-defined events as well as duties of financial contribution' (Ferrera, 1996: 49).

With regard to families, while policies act on them, policies in their turn are also affected by the formation and behaviour of families themselves (Fox Harding, 1996). This means looking at the twin interaction between the two elements of the analysis: policies and families.

The Relationship between the Family and the State: A Theoretical Perspective

The relationship between the state and the family has always been crucial for understanding the material and symbolic organization of families. The fundamental aspect of this relationship is the institutional dimension of 'the family', the fact that the family is one of the main institutions which provides for social reproduction and social control in society. The institutional dimension of the family is the real object of state intervention and it is within this relation that we experience the different organizational forms of what we label 'family'. From an analytical perspective, the institutional dimension of the family is assumed to be a metaphor for the opposing dichotomy between the public and the private spheres. The nexus between public and private exists only in terms of this relationship. As a consequence, the concept of private does not have life of its own without that of public. Thus, the modern family exists only within defined institutional expectations of the society and as a consequence of the state (Saraceno, 1984).

The relationship between the family and the welfare state in the area of social reproduction is where interference between the public and the private spheres is strong.[1] The state – by its welfare state structures – has been by no means neutral with regard to the family. A range of measures, from transfer payments to welfare provisions, have been grounded on certain presuppositions about family functioning and have attempted to create, strengthen or modify the appropriate conditions.

In Moroney's words, the development of the welfare state during this century has been based on:

> A set of implicit and explicit assumptions concerning the responsibility which families assume, or are excepted to assume, for their members and the condition under which this responsibility must either be shared with or taken over by society through its public or voluntary organisations. (Moroney, 1976: 5)

The historiography of welfare states has tended to focus almost exclusively on the role of the state and/or on the nexus between the state and the market, neglecting the importance of other relations. As Lewis suggests, it is more adequate to see modern states as always having had a 'mixed economy of welfare', in which the state, the voluntary sector, the family and the market have played different roles at different points in time. Although the mixed economy of welfare has taken different forms in different European countries, the family has always been the main provider of welfare in every country (Lewis, 1993). Paci (1987) has also noted that the major challenge to comparative work on the history of welfare regimes is to chart and explain the changing boundaries between the various elements in the 'mixed economy'.

From this point of view the issue of family contribution, particularly the contribution of women to the welfare state and to social protection, underlines the importance of considering private issues in the design of public protection measures. It highlights the need to take into account not only what happens with respect to access to work and to work's accompanying protection systems, but also what happens in family life and the social protection that it provides (Martin, 1996). This means that family issues should no longer be treated as sectoral issues, but rather in terms of family policy or interventions undertaken by the state or other social partners for their support.

The balance of responsibilities between the family and the state in the provision of social welfare is a key issue in social policy analysis. The boundary between public and private spheres has been redrawn, over time, at the level of prescription as found in certain laws and social policies. One way of looking at state policies about the family is that these are in effect concerned with drawing the boundaries between state and family and defining the proper role of each. It should be added that the process of drawing the boundaries between state and family responsibilities is part of an ongoing process and is a contested arena (Finch, 1989).

Thus, not only must the division of labour between different institutions be taken into account in social policy analysis, but also the gender division of labour and the forms of interdependence between men and women and between parents and children within families (Saraceno, 1994; Sainsbury, 1994b). As a consequence, cross-national variation between welfare states (e.g. Esping-Andersen's work[2]) and in family patterns cannot be understood without looking at the ways in which welfare state policies have conceptualized paid and unpaid work.

The Convergence of Research on the Sociology of the Family and Welfare State Studies

This section illustrates the fruitful intersection of work from sociology on the family with studies on the welfare state. An overview of the existing literature leads to the conclusion that sociologists of the family were the first to pay attention to the relationship between social policy and the family. It was only with the awareness of the limits of welfare state growth that scholars on the welfare state began to re-evaluate the family in their analyses of welfare state functioning.

The Sociology of the Family and Interest in the Welfare State

In the area of sociology of the family, there are two very different approaches to the study of the interrelationship between the family and the welfare state in the European context:

1. The feminist approach, namely, gender studies and studies on caring work.[3]
2. The approach that conceived the development of the welfare state as an example of excessive state intrusion into the family's private sphere.

Since the 1970s gender studies and research on caring work have begun to underline the importance of studying the relationship between women, the family, and the welfare state. The authors who adopt this approach conceive of the family as an institution reacting to state intervention as well as influencing other social institutions. This means that the family cannot be seen merely as a 'dependent' variable. Instead it should be conceived as a complex institution, composed of different, and sometimes conflicting, individuals with different needs, interests and perspectives. This perspective also emphasizes that the family changes its forms and functions across social, historical, economic and cultural contexts.

From the 1970s onwards, feminist analyses of the family and the welfare state have focused on the social control exercised over women by various public bodies for the organization of reproduction (Pitch, 1983). In particular, studies of how welfare states organize social reproduction describe the relationship between women and the welfare state in terms of both 'patriarchy' (e.g. Eisenstein, 1979) and 'partnership' (e.g. Siim, 1987). McIntosh (1978), amongst others, argues that state policies contribute to the reproduction of gender inequalities and of 'patriarchy'. She approaches the question from the standpoint of how state intervention acts to maintain the family household and waged labour – two institutions which she sees as oppressive to women. In her analysis, the state acts through its support for a particular form of family in which women provide unpaid domestic work.[4] This means that in part of the international literature, the assignment of caring to women has often been assumed to contribute to the structurally inferior situation of women in the modern welfare state (Leira, 1992).

Other feminist scholars have emphasized the idea that welfare states work in order to ameliorate social inequalities. In other words, they describe the relationship between women and the welfare state in terms of a 'partnership'. Feminist scholars holding this view come from backgrounds that focus on gender, poverty and class inequality. They argue that the high poverty risks for women can be reduced by generous income transfer programmes (Orloff, 1996). This understanding of gender relations is associated with the Scandinavian research tradition, which offers a more positive interpretation of the relationship between women, the family and the welfare state. This tradition, for instance, speaks in terms of a 'women-friendly state' (Hernes, 1989; Lewis and Åström, 1992).[5]

Since the end of the 1970s a specific and very rich feminist school of thought concerning the relationship between the welfare state and the family has also been visible in the Italian context. This research is concerned with the crucial importance of household work, as well as the presence of women in the social services labour market, for the adequate functioning of the welfare state (Balbo, 1978; Bianchi, 1981; Saraceno, 1984).[6] The scholars adhering to this school of thought argue that with the development of the welfare state, families have not lost their function, as functionalist theory maintains, but, on the contrary, have been burdened with new tasks and responsibilities. These feminist scholars maintain that the use of any social services requires some family work. Thus, the welfare state develops because women are at the centre of the social networks that connect reproduction within the family to the services provided by public agencies. In the Italian context, this area of research has also shown that the development of the welfare state has encouraged the creation of a specific kind of family, organized on a clear division of work and allocation of time between the sexes. This phenomenon has been labelled 'double presence', and it has led to a new model of gender identity that goes beyond a woman's role in the family to establish a new equilibrium between public and private life (Balbo, 1978; Bimbi, 1989).[7]

The authors belonging to the second approach were interested, as previously mentioned, in studying the effects the development of the welfare state has on family private life. In contrast with the feminist perspective, these authors have a vision of the family as a unit, composed of individuals with homogeneous preferences and interests. Although there are important differences within this approach – for example among Lasch (1977), Donzelot (1977) and Donati (1981) – all the authors formulate a variant of the Parsonian idea of the diminishing functions of the family in modern society and, in particular, as a consequence of the development of the welfare state. The main reason for clustering these authors together is that all of them share a vision of the state as an authoritarian institution which aims to destroy and invade the private space of the family. The main focus of these authors is the effect that the development of the welfare state – and in particular of social services for children and the elderly – has on the functioning of the family unit. In Italy this strand of research has been particularly important and has been adopted mainly by Catholic writers (Ardigò, 1974; Donati, 1981, 1984). They opposed the spreading of public facilities for young children because they were seen as limiting or reducing the educational independence of the family (Donati, 1989).

Welfare State Studies and the Family

As far as welfare state studies are concerned, the family has been viewed from two different perspectives:

1. As an important provider of services, that is, as an important component of the 'welfare society' or 'welfare mix', especially in phases when growing service demands are not likely to be met with increases in public expenditure.
2. As having lost many of its traditional functions due to the effects of modernization processes. As a consequence, family changes are seen as important elements for understanding the development of the welfare state and how it has varied across countries.

The re-evaluation of the family in welfare state studies was mainly the logical consequence of the socio-economic crisis that has affected all European countries since the mid-1970s. Since then, for most European countries the years of confident growth and welfare expansion have been succeeded by a collective loss of faith in the possibility of uninterrupted growth of the economy and disillusionment concerning the programmes and promises of a welfare society. The growing awareness of the limits of the welfare state has led to a reappraisal of the role of the market, the family, and voluntary organizations (institutions regarded as alternative mechanisms of resource allocation for the provision of social protection and welfare) (Paci, 1987).

In the new economic context, the social division of responsibilities among various institutional spheres – the state, the market, and the family – has had to be renegotiated (Sgritta, 1989). As Flora points out, the essential meaning of the changes produced by the crisis 'may be found much less in the threat of an historical regression of the European welfare state, and much more in the break of the past growth patterns which will require a readjustment of major institutions' (Flora, 1985: 25).

In Flora and Heidenheimer's view, the welfare state is to be seen as an institution that made 'consonant with modernity many functions previously performed by the family, the church, the guild, and local community' (Flora and Heidenheimer, 1981a: 6). As a consequence, the welfare state develops where there is modernization and where other 'traditional' institutions and associations have become weak.

Comparative Analysis of Welfare States

The Development of the European Welfare State

The 'launching' of the modern welfare state originated in the nineteenth century, in the period ranging roughly from the unification of Italy and

Germany to the First World War. Despite the common impetus for European welfare states, there are many variations and sources of diversity among them. These date back to phases in their development which often pre-date the First World War. Early decisions were difficult to reverse, especially when taken under exceptional circumstances, for example in times of war. Following Flora (1986), two basic dimensions of the institutional infrastructure of the welfare state are important for understanding their future institutional characteristics: (1) the degree of *stateness,* or the degree of state penetration into welfare institutions and the degree to which the instruments of government are differentiated from other organizations (in other words, the degree of penetration of the state *vis-à-vis* the development of the various non-state 'intermediary structures'); and (2) the degree to which welfare institutions reflect social differentiation, that is, the degree of fragmentation of the welfare state.

One of the main historical points of differentiation of the state penetration of welfare institutions is related to the presence and strength of other civil society associations. Thus, according to some analysts, the Church, or more precisely the variation in the relationship between the Church and the state since the Reformation, has to be considered an important variable in explaining welfare state variations. The Catholic Church generally has emphasized the principle of 'subsidiarity'. During periods of strong influence it has done much to hamper the development of public child care services and public social assistance policies and services.

From a historical point of view, the development of European welfare states can be separated into four periods: (1) the take-off period, from about 1890 to 1920; (2) enlargement, from 1920 to 1945; (3) consolidation and institutionalization, from 1945 to 1970; and (4) crisis, from 1970 onwards. By 1950, most western European countries had established a social security system which protected workers in the formal labour market and their families against four basic risks of life: old age, illness, unemployment and disability. Three western European countries constituted an exception to this: Greece, Portugal and Spain (Flora and Heidenheimer, 1981a).

The Development of Comparative Welfare State Research

For some time, comparative research on welfare states has been divided into two main approaches: narrative historical monographs, on the one hand; and sweeping statistical comparisons, consisting of identifying and measuring differences and similarities in welfare states by quantitative indices, on the other (cf. Ferrera, 1993). More complete analyses of the institutional profiles of various welfare states have been developed only recently (e.g. Esping-Andersen, 1990; Orloff, 1993; Pierson, 1994; Baldwin, 1990).

Jens Alber (1995) distinguishes three main phases in the development of comparative welfare state research:

1. Up to the late 1970s: the leading paradigm in welfare state research in this period was that of functionalism. An example of this type of approach is offered by Wilensky, who, by comparing social security spending as a proportion of GDP, arrived at the conclusion that cross-national variations can be explained by the number of problem pressures and the amount of available resources of a country (Wilensky, 1975).
2. From the late 1970s to the early 1980s: in this period conflict theoretical approaches were dominant. Authors such as Korpi (1978) in Sweden and Castles (1978) in England are the main representatives of this approach. According to these authors, the differences in social politics of the countries are the main explanatory factors for welfare state variations. The variation may best be explained by looking at the balance of power between organized labour and liberal/conservative forces.
3. Since the mid-1980s: during this period the school of institutionalist analysis became prominent. The authors belonging to this school have pointed out the important role played by states and their bureaucracies in shaping and reforming social policies. The pioneer of this school was Heclo (1974), and later Skocpol (1985) and her colleagues, who refocused the research agenda during the 1980s on the roles played by political elites and institutions.

More recently, early attempts to find out which specific variable might explain cross-national variations in social policies have given life to *configurational analysis*, which identifies the constellation of factors that explains the development of welfare states (Ferrera, 1993). The first attempts in this direction were already made in the 1970s, by authors like Rimlinger (1971) in the USA and Flora and Heidenheimer in Germany (Flora and Heidenheimer, 1981a). As Alber points out (1995, p. 133), following in their footsteps, we are now witnessing new institutional arrangements, but in outcomes rather than outputs of public policy.

The Explanations of Welfare States

From a theoretical perspective the development of welfare states within the different approaches and the different historical phases mentioned above can be divided into two general groups, depending on the degree of autonomy given to the state in shaping social policies.

'Demand-based' theories The first group can be defined as *demand-based* or *society-centred* theories. These attempt to explain the development of the

welfare state in terms of the demands for policies posed by different groups or social classes. From this perspective social policies are the answer to social demands articulated by interest groups and political parties. There is a wide range of theories within this group of explanations. Firstly, the previously mentioned *functionalist theories* (Wilensky, 1975; Rimlinger, 1971). In opposition to these functionalist theories, an alternative interpretation focuses on how partisan politics contribute to cross-national differences among welfare state institutions. These interpretations try to identify the interests served by social policy, as well as the key actors in the introduction of such policies, emphasizing the political struggles that have underpinned the development of the welfare state. I previously classified these explanatory theories as *conflict theoretical approaches*. Within them the *politics matters* perspective and *power resources* school of analysis[8] assume that class division between capitalists and wage-earners in industrialized capitalist democracies drives the development of welfare states (Korpi, 1978; Orloff, 1993).

'State-centered' theories These theories maintain that states can affect the development of social policies autonomous of external demands or the interests of social groups (Skocpol and Amenta, 1986). A key concept within these explanations is the concept of *state autonomy*. States are not conceived as merely arenas of political conflict or as passive administrative tools to achieve the demands of the most influential social and political groups in the policy-making process. They are organizations with autonomous capacities to carry out independent actions. Within this perspective, the process of state-building and state structures themselves not only affect state capacities, but also the political struggles through which social policies have evolved, influencing possibilities for collective action and political alliances. The study by Weir, Orloff and Skocpol (1988) identifies past policies as shaping policy-making through effects ('feedback')[9] on issue agendas, policy discourses, political coalitions and the models of institutional development.

Different Types of Welfare State

An influential contribution to recent comparative welfare state research – and a work which has stimulated wide debate – is Esping-Andersen's *The Three Worlds of Welfare Capitalism* (1990). In this, the author argues that western welfare states may be classified into three regimes, which he labels 'liberal', 'conservative-corporatist' and 'social democratic'. The starting point of Esping-Andersen's classification is Titmuss' well-known trichotomous categorization of welfare states based on social rights and the conditions for entitlement to benefits (Titmuss breaks these down on the basis of need, work performance, and universal rights). The 'liberal' type of regime can be found in the Anglo-Saxon countries, the 'conservative' in continental countries (Austria, Germany, France Belgium and Italy) and the 'social democratic' in the Scandinavian ones. Esping-Andersen's typology is based on three dimensions: 'state–market' relations, 'stratification' and 'decommodification'. The state–market relation emerges from the specific institutional configuration of market, state and family adopted by societies in the pursuit of work and welfare (although the relation between family and state is neglected in the analysis).

The key issues in determining the variation between welfare states are measured by the social impact of policies and in particular by concepts of 'decommodification' and 'stratification'. The former is strictly linked to the concept of social citizenship, and its different degree across countries demonstrates variation in social rights and in how people are permitted 'to make their living standards independent of pure market forces' (Esping-Andersen, 1990: 3). The latter is used to measure the degree to which public benefits tend to segment or integrate populations. According to Esping-Andersen, while social policy is supposed to address problems of stratification, it also produces them. The author's question is 'does the welfare state create dualism, individualism, or broad social solidarity?' (p. 4). Feminist scholars have criticized Esping-Andersen's *Three Worlds of Welfare Capitalism* (Orloff, 1993; Lewis and Ostner, 1994; Bussemaker and Van Kersbergen, 1994; Daly, 1994) mainly because the nexus between the family and the state is not taken into account in his typology. In the empirical analysis the family is mentioned only in the analysis of the conservative regime (although not in a disaggregated way – in the sense of the gendered and generational divisions of labour), while hardly appearing at all in discussions of the other two regimes. However, it should be noted that in recent years Esping-Andersen has acknowledged his lack of systematic attention to the family in this book. In his most recent work he revised his analysis on welfare state regimes through the lens of the family, although exploring gender differences was not a goal of the analysis (Esping-Andersen, 1996, 1999). In the first work (1996), he suggests that the US model, with its large, low-cost service sector and patterns of high mobility throughout the life course and across the employment sector, could provide answers to two of the main western European economic and social problems: high unemployment and low fertility rates. According to the author, the best solution to these problems is to foster 'post-industrial' families, especially important in countries such as Spain and Italy, where levels of unpaid (family) work remain high. In his latest book (1999) a whole chapter is devoted to the nexus between families and welfare regimes. In this work he revises his previous analysis, responding to feminist criticism and recognizing the actual welfare role assigned to families in different welfare states. In any case, the main aim of the work is to examine the degree to which families absorb social risks, and the focus of the analysis remains the family rather than gender relations, and/or intergenerational relationships.

A different welfare state typology has been provided by Ferrera (1993), who distinguishes variations in welfare states on the basis of the model of coverage adopted by different countries.[10] Historically, in the author's view, the model of coverage measured by what has been considered as the core of the welfare state, that is, social insurance, has had two fundamental variations: (1) the 'occupational model' (based on work performance); and (2) the 'universalistic model' (based on citizenship principles). As the author clarifies, these two clusters have only classificatory purposes, as sorts of 'polar types'. Ferrara differentiates the occupational and universalistic models into four sub-categories, as indicated in Table 1.

Table 1
FERRERA'S TYPOLOGY OF THE WELFARE STATE

Occupational model*	**Universalistic model**
Occupational pure	*Universalistic pure*
France, Belgium, Germany, Austria	Finland, Denmark, Norway, Sweden
Occupational mix	*Universalistic mix*
Switzerland, Italy, Netherlands, Ireland	New Zealand, Canada, Great Britain

Note: * USA and Australia are considered as two cases *sui generis*.

Each country is pigeon-holed into one of the four categories. Inside the occupational model, which is most interesting for the purposes of this work, a distinction is made between the occupational pure model and the occupational mix, which includes those countries that have deviated in some aspects from the original occupational form (e.g. Italy, for its health system, since this approaches the universalistic type).

Towards New Directions of Comparative Research

The 1990s have corresponded with a season of great progress for comparative welfare state research, in particular the emergence of two new strands of research, important both for the explanations of historical variations of welfare state systems and for a wider comprehension of the development of southern welfare states. In the first strand all the comparative welfare state studies that have adopted a gender-sensitive approach can be included. The second comprises all the studies which have contributed to illuminating the existence of the Mediterranean welfare state model.

Gendering welfare states in comparative perspective It was only in the 1990s that a number of analysts began to develop analytical frameworks suitable for comparative work on the relationship between gender and social policies. The strand of gendering welfare states (of the 1990s) emerged as a criticism of mainstream theories and conceptions in comparative welfare state research. The introduction of a gender perspective has scrutinized several basic concepts in the mainstream literature by inquiring how they are gendered. Indeed, most of these scholars have addressed their criticism not generically to the mainstream theories but to the well-known Esping-Andersen typology.

Most scholars (Lewis, 1992, 1993; Orloff, 1993) point out, for instance, that the concepts used in mainstream comparative welfare state research, such as 'decommodification' or 'dependency', have a gendered meaning. These concepts have different consequences when applied to women and to men, consequences that are rarely acknowledged in welfare state analysis. For instance, while Esping-Andersen argues that decommodification is a necessary prerequisite for workers' political mobilization, the worker he has in mind is a man whose political mobilization may depend on unpaid female household labour. For women, decommodification is likely to result in their carrying out unpaid caring work (Lewis, 1992). In addition, most scholars have argued that it is necessary to develop a conceptual framework for analysing the gender content of social provision. First of all, the state–market relations have to be extended to consider the way in which countries organize the provision of welfare through families as well as through the market and the state (Bussmaker and Kersbergen, 1994; Daly, 1994; Orloff, 1993). Second, what is needed in mainstream theory is a concept of the welfare state that incorporates the relationship between unpaid work, paid work and welfare. In other words, scholars need to recognize that the private–domestic division is crucial for any understanding of women's position, because, historically, women have typically gained welfare entitlements by virtue of their dependency status within the family as wives (Lewis, 1992). Taking into account these considerations and the necessity of analysing the informal private sector (welfare provisions provided by the voluntary sector and the family), Lewis and Ostner (1994) provide the foundation for an alternative conceptualization and typology of the welfare state. They argue that 'the majority of modern welfare states may be categorized as exemplars of a *strong male breadwinner* model. In its ideal form, this model prescribes breadwinning for men and home-making/caring for women' (ibid., p. 17). According to the authors, these assumptions were built into welfare policy provisions to different degrees in different countries. This has produced varying results in terms of *women's entitlement to benefits*; *levels of social wage*; *public expenditure for services*; and *female labour participation*. In particular, Ostner and Lewis contrast three types of countries:

1. Countries where a *strong male breadwinner* model prevails (Germany, the United Kingdom and the Netherlands). In these countries, women's social rights are almost exclusively derived rights attained by virtue of their status as wives. Women are more likely to receive 'second-class' assistance benefits rather than 'first-class' insurance benefits. Here, states do little to encourage the presence of women in the labour market and they tend to promote policies that reinforce the principle of obligation within the household.
2. Countries with a *moderate male breadwinner* model (France). Here the state recognizes and promotes women's position as both worker and mother. The state has developed a 'parental model' of the welfare state, providing more extensive and explicit public provisions for dependent children. However, in the case of France this model was primarily the product of both pro-natalist policies and social Catholicism (Schultheis, 1992).
3. Countries with a *weak male breadwinner* or *dual-earner* model (Sweden, Norway and Finland). In these countries women have been defined as workers rather than as wives and mothers since the 1970s. These welfare states promote 'two-breadwinner families' (Lewis and Ostner, 1994).

This typology has also been widely discussed and criticized, by the same authors,[11] opening a wider debate among feminists concerning the most appropriate criterion and the most suitable social measures to be analysed in order to introduce a gender dimension into welfare state analysis and typologies.[12] Various authors (Hobson, 1990; Orloff, 1993; McLaughlin and Glendinning, 1994) have proposed that a useful criterion for building a gender-sensitive typology of welfare states may be to consider the degree to which it is possible for women, as well as for men, to form and maintain an autonomous household. These authors, recalling the concept of decommodification, suggest replacing it with concepts such as 'personal autonomy' or 'defamiliarization'. From this perspective defamiliarization would indicate 'the terms and conditions under which people are engaged in families, and the extent to which they can uphold an acceptable standard of living independently of the (patriarchal) family' (McLaughlin and Glendinning, 1994: 65).

Gendering the historical research An important strand of research into gendering comparative welfare state studies emerged in the 1990s from theoretically oriented historical research that focused on the long-term development of state provisions and their gendered effects.[13] Four historical works have been especially important. These include two edited volumes (Koven and Michel, 1993; and Bock and Thane, 1991) and two comparative studies of Britain and France (Jenson, 1986; and Pedersen, 1993). The two collections contain many single-country case studies, placed in comparative perspective, on 'maternalist'[14] political activities and social policies from the nineteenth century until the mid-1950s (Orloff, 1996; O'Connor *et al.*, 1999).

The main findings of these recent collections are that although there were impressive similarities in maternalist politics and policies towards women and children among countries, there were also important variations in women's political presence in policy-making as well as in policy outcomes relevant for women and children. According to Koven and Michel there were differences in outcomes between 'strong' and 'weak' states. While women's movements and women's involvement were stronger in the so-called weak states (i.e. Britain and the USA), policies aimed at protecting children and women were better developed in the so-called strong states (i.e. Germany and France).

Bock and Thane focus their work on seven countries in western Europe. Their collection explores the relationship between the parallel development of welfare states and of women's movements between the 1880s and 1950s. They illustrate the varying ideas about gender that these processes both embody and construct. By concentrating on the first measures concerned with maternity, the collection sheds light on the importance and the extent of early maternal welfare provisions in the development of welfare states. In addition, these case studies highlight the importance of women's contributions to the construction of welfare states. According to the authors, feminist discourse at that time focused on maternity and the improvement of the lot of mothers in a sort of transnational movement. However, there were important variations across different countries, particularly at the level of concrete proposals and policies. For instance, there were wide differences in policy outcomes between countries with democratic regimes in the 1930s and 1940s and those that had fascist experiences. In France, Britain, Norway and Sweden, for example, feminists throughout the interwar period demanded universal maternity or child allowances regardless of women's occupational status, and state child allowances have been payable directly to mothers from the time of their introduction or soon after. This seems not to have been the case in Italy, Germany and Spain. Although turn-of-the-century feminists in Italy and Germany asked for child allowances for all mothers, they soon devoted their attention towards maternity benefits for employed mothers. Paying family allowances to families with non-employed mothers and over the whole period of child-raising seems to have been a state initiative rather than a woman-led initiative. It is thus not surprising that in Fascist Italy, Franco's Spain and National Socialist Germany, child allowances were paid to the fathers.

Two other recent historical and comparative works have been particularly influential in understanding the mutual effects of state social policies and gender relations and the interactions among class, gender and other social relations. The first is Jenson's (1986) comparison of Britain and France, and the second is Pedersen's (1993) study of the same two countries.

According to Jenson's work, both French and British political elites made policy decisions in an international context where population concerns, particularly those regarding declining birth rates, were at the forefront. However, these two countries did not respond to this problem with similar strategies. The author shows that differences in the capacities of organized workers and trade unions, different levels of demand for female employment, and different discourses and ideologies about motherhood and paid work produced strikingly different policy outcomes. The British government worked to make support for children possible through fathers' wages alone (the family wage), while the French government encouraged families to have two incomes. French women were assisted as mothers and workers, receiving both economic and health-related services for themselves and their children.

Pedersen's (1993) comparison of British and French policies shows the emergence of state policies towards family dependence; she attempts to understand why these welfare states developed along such different gender lines. Pedersen concludes that in Britain, both the labour market and the social welfare system developed along male breadwinner lines, disproportionately distributing benefits to men on the assumption that they were supporting dependent wives. In France, both wages and welfare policies became more 'parental', supporting mainly families with dependent children. According to the author, the balance of power among unions, employers, and the state, coupled with the types of discourses about mothers' paid and unpaid work, were the most influential variables in determining policies *vis-à-vis* dependent children and women's labour force participation.

The most important contributions that have been provided by these historical studies is that they have focused on the longitudinal effects and variations of state provisions on gender, and also that they have been able to show the importance of focusing on the historical evolution of policies towards families, women and children.

Mediterranean welfare states[15] As far as comparative welfare state research is concerned, the academic debates have only very recently begun to systematically include the Mediterranean countries within their scope of observation (Greece, Italy, Portugal and Spain, with some doubts as to whether to include France or not). The issue is whether systems of social protection in these countries have a logical distinctness from other European countries, or whether they are simply less developed. With the exception of France and Italy, the other Mediterranean countries have been neglected in comparative welfare state studies. Both Titmuss and Esping-Andersen included Italy and France in their studies, with Titmuss subsuming them into his 'industrial achievement' category while Esping-Andersen groups them into his 'conservative–corporatist' regime, with Germany, Austria, and Belgium. In the few examples in which Spain, Portugal and Greece have been taken into account, they have been treated as late-comers (Ferrera, 1996; Castles, 1994).

The first author who treated the 'Mediterranean' countries as a distinctive cluster was Leibfried (1992). According to this author, 'Latin rim' countries, which include Spain, Portugal, Greece, Italy (to some extent), and, to a lesser extent, France, are characterized by a 'rudimentary welfare state'. In some respects, these countries are similar to the Anglo-Saxon countries, stressing 'residualism' and forced entry into the labour market. However, in this part of Europe older traditions of welfare (related to the Catholic Church) are recognized as being very important. In addition, in some Mediterranean countries certain social security programmes serve as basic income measures – although they were not designed to do so (e.g. the disability pension in southern Italy). Moreover, labour market structures are different, often revealing a strong agricultural bias combined with 'subsistence' economies. Finally, southern European countries do not have full-employment traditions, as do some of the other European countries. Other authors, such as Castles (1994), have pointed out the influence of Catholicism and the important role its social doctrine plays in shaping the welfare state in these countries. Along these lines, Van Kersbergen (1995) has tried to answer the question of whether Christian democracy fosters a distinctive welfare state regime or not. According to this author, the first answer is that 'it is not so much politics or politically generated institutional settings that matter in explaining cross-national qualitative differences between welfare state regimes, but rather it is cultural variables that are of consequence' (p. 229).

In general, the Catholic Church has continued to emphasize the principle of 'subsidiarity', by which is meant that priority is given to smaller groups and voluntary organizations over the state, whenever it is possible. According to Spicker (1991), this concept suggests that decision and action should always be taken at the lowest level possible. The author, by examining the relationship between subsidiarity and the family, maintains that

> Subsidiarity is justified as an expression of the responsibilities that people have for each other's welfare. These depend on the closeness of their relationship. Because the closest relationship mainly exists within families, it is the family that bears the primary responsibility for social support. The role of others who are more remote is correspondingly reduced; they are 'subsidiary' to the primary responsibilities. The role of public services, in particular, is subsidiary to that of family, local community and the private sector. (Spicker, 1991: 4)

In this regard, particular attention should be paid to the development of care services (for both the elderly and children) in Mediterranean countries, because of the traditional importance of the Church in this field, and to the development of public poor relief policies and the influence of the private, religious groups. However, the principle of subsidiarity does not necessarily mean that there is no role for government action; it relates more to the nature of the interventions made.

The idea that Mediterranean countries (Italy, Greece, Portugal and Spain)[16] compose a distinctive cluster with its own characteristics and experience has been widely developed by Ferrera (1996). According to Ferrera, the southern European countries are characterized by a unique mode of political functioning. In the 1980s social scientists coined the expression 'particularistic-clientelistic' to describe a main feature of the Italian welfare state. From Ferrera's point of view, the primary distinctive characteristics of the southern European welfare state are

1. High transfer payments (especially pensions). As in the other 'Bismarckian' and 'corporatist' countries, southern European income maintenance is based on occupational status, and its degree of institutional fragmentation is very high. From Ferrera's point of view, these countries have implemented a two-fold, 'dualistic' protection system which provides high coverage for whoever works in stable employment, and very weak coverage for those who do not. No national income scheme exists, and there is unbalanced distribution of risks between generations and across regions. In these countries we find an underdevelopment of family benefits and state housing policy.
2. National health care: (near) universalism, with extended scope for private provision. This means that health-care systems in these countries, contrary to those in other 'corporatist' countries, are related (at least formally) to 'citizen's rights' rather than to 'work status'. In addition, in southern European countries health-care coverage is defined by a private–public mix. The establishment of national health services has

created a particular 'collusion' of public and private – often with great advantages and profits for the private sector (Ferrera, 1996: 13).

3. Low penetration of the welfare state and low degree of 'stateness'. The first characteristic refers to the permanence of the private–public mix, as mentioned in the case of health care. The second characteristic refers to the low degree of state autonomy, with a persistence of partisan ties, clientelism and the existence of a patronage machine. Access to social welfare even comes close to being a relation of monetary exchange between political and social groups, including trade unions, and particularly at the local level.

According to Ferrera, explanations of the 'southern road' emerge from two different perspectives. The developmental perspective explains the emergence of social policy in these countries by pointing to their specific paths to modernization, which have been characterized by entrenched 'backwardness'; market, sectoral, and territorial dualisms; amoral 'familialism' and civic parochialism; the strong presence of the Catholic Church; and the exclusion of left-wing parties from government. From a politico-institutional perspective, the southern European welfare state is interpreted primarily in the light of the 'power game' resulting from the specific structure of Mediterranean polities. The specific traits of polities in southern European countries include a weakness of state institutions, the pre-eminence of parties for the aggregation of social interests, and ideological polarization (ibid.).

Finally, Moreno (1997: 2) emphasizes the 'cultural-axiological' dimension of the southern European model of the welfare state. Moreno argues that these countries share analogies in their value systems, such as 'a self-perception of differentiated needs and lifestyles', intrafamilial pooling of resources and home ownership, 'a compelling household solidarity', and the prevalence of values of family inclusion and intergenerational redistribution (i.e. gift mechanisms, and diffusion of family enterprises).

A Model for Analysing Social Policy in Mediterranean Countries

Resources and Limits of the Existing Theoretical Frameworks

Mediterranean welfare states cannot be completely understood if they are included in the 'conservative' or 'corporatist' regimes (as in Esping-Andersen's 1990 study). In this cluster of countries the cultural emphasis on the family is not translated into high public economic transfers. At the same time, to interpret social policy in Mediterranean countries merely as 'rudimentary' prevents us from understanding the specific historical configurations that have affected social policy and the role of the family in the south. In this direction, the strand of research that has developed the idea of a Mediterranean model of the welfare state has contributed both to advancing the acknowledgement of the historical characteristics of southern countries and to a better understanding of the welfare state theory.

The introduction of a gender perspective into comparative welfare state research has shed light upon variations in the effect of policies on different family roles (gender roles) and on the distinct underpinning of social rights for men and women. However, this approach has not taken into account the complexity of relations and meanings related to the experience of families. Particularly in southern European countries, family experiences and family definitions, both in terms of relations and of structures, cannot be conceptualized adequately only in terms of household structures or nuclear families. While gender analyses of the welfare state have framed the family as an institution divided mainly by gender, it has failed in assuming a wider perspective capable of including other fundamental family divisions: generations and kinship.

A New Theoretical Perspective: Family Models

Does a convergence of these two strands of research exist, which can contribute to a better understanding of the functioning of Mediterranean welfare states as well as to advance current theories on the welfare state?

First, a theoretical perspective on social policies that takes into account the relationships between the state, the market, and the family within the specific historical context of southern European countries is needed. Second, it is necessary to introduce family and kin relations, from a gender and intergenerational perspective, into social policy analysis. In particular, it becomes crucial to detect how public and private obligations and responsibilities were legally defined in legislation and social policy with regard to the needs of economically *dependent* individuals on the basis of age, generation, gender and degree of kinship.

This marks an important departure from theories that emphasize the concept of the 'male breadwinner' family model; that perspective neglects both how social policy has treated children in relation to the elderly, as well as how kinship obligations extend beyond the nuclear family. Indeed, it seems to me that we need to take into account not only gender issues but, more broadly, the ways in which the issue of family responsibilities and that of dependent family relationships are acknowledged, framed and defined at the institutional level, in social policies, fiscal policies and laws.

Studies showing that important variations exist across countries in terms of family obligations have also moved towards introducing a focus on family and kin relations (Millar and Warman, 1996). The need to focus on the content of social policies, in terms of family and intergenerational links and obligations, and especially the need to distinguish between parent–children and children–elderly parent obligations, have risen in comparative analysis of social care services addressed to children and to the elderly (Anttonen and Sipiliä, 1996; Lewis, 1998; Saraceno, 2000). These works have acknowledged the importance of introducing family relations fully into the analysis, though not always at the theoretical level, from a gender and intergenerational perspective, mainly interpreting intergenerational and kin ties as residual relations, doomed to disappear.

Bearing in mind the theoretical contributions of recent research on Mediterranean welfare states and in gender studies, this book aims to bring the family fully into welfare state analysis. The nexus between the state and the family is investigated here by looking at how social policies, fiscal policies, and family law have codified the division of responsibilities between the state and the family in the areas of reproduction and caring work. In general, we can assume that there are two main types of obligations defined in the area of reproduction and care:

1. Financial obligations, which define who should financially support whom, and which is a fundamental aspect of income maintenance policy.
2. Caring obligations, which prescribe who should provide care for whom, and which comprise issues that are central to social and health provision. (Millar and Warman, 1996)

The division of responsibilities between the state and the family might also be seen in terms of the boundary between public and private responsibilities, which has been redrawn over time at the level of prescription as found in certain laws and social policies, and which has resulted in different institutional definitions of family models.

Of course, in analysing the different definitions of family models, a particular emphasis is placed on the normative aspects embodied in legal and social policies. I argue that social policies and laws develop on the basis of a set of assumptions about how families function in society, and take for granted certain gender and generational relationships within the household and between kin. At the same time these laws and policies play an active part in creating social norms about appropriate roles, activities and behaviour.

How laws and social policies have defined the family How have social policies within European welfare state systems promoted or encouraged ideal family types? Which types of family models have been implicitly supported by social policy? Is it possible to build a typology of family models in which, from a gender and intergenerational perspective, the prescription as found in certain laws and social policies comes out?

As a heuristic exercise, in table 2 a typology of 'ideal' family models is shown. This has the purpose of illustrating how certain laws and social policies have contributed to creating different definitions of the family and in this way governed gender, generations and kin relationships. To this end certain crucial dimensions of analysis have been privileged:

1. Assumptions about how the family actually works (e.g. interaction between the structure of families and the characteristics of the labour market; the division of gender roles in the family; the strength and intensity of relationships, and so on).
2. Normative statements about how the family should work.
3. Policies and laws explicitly designed to support the desired functioning of the family and its members.
4. Policies and laws that indirectly or implicitly seem to support a certain model of the family.

It should be kept in mind that the characteristics of the models presented in Table 2 do not exist in a 'pure' form in any country; they are in fact presented as contrasting 'ideal types'.

I have distinguished three main family models. The first model, the 'male breadwinner model', prescribes breadwinning for men and home-making for women. In general, this model should be found in a 'pure' occupational welfare state model (cf. Ferrera's typology). Here family relations tend to be bound within the nuclear family, both at the level of prescriptions and at the level of family behaviour. The assumption is that the wife and the children are financially dependent on the husband/father. In other words, the principles of breadwinning heads of household and a family wage underlie law and social security insurance policies. These serve in many cases as the basis of recognition for wives' and childrens' derived entitlements (e.g. survivors' pension, health insurance) and for the higher benefits the husband/father receives by way of dependants' allowances (e.g. family allowances, tax allowances, unemployment supplementary allowances, etc.). The principle of breadwinner is further supported by the provision that men who survive their spouses receive different treatment than women in the same situation. Finally, a scant development of child care services and services for the elderly is found, given that this work is considered women's work.

The second model, the 'dual-earner' model, should be found in countries whose welfare states are mainly grounded on the principle of citizenship (cf. the 'universal model' in Ferrera's typology). This model assumes full employment for both men and women, thus extending individual rights, such as pension rights, to everyone in the household. Children, although defined as dependent on their parents until they come of age, are also entitled to social rights. Moreover, social services for children and for the elderly are expected to be developed in order to enable women to combine paid work and motherhood.

The third model, the 'family/kinship solidarity' model, should be related to Ferrera's 'occupational mix' welfare state model. As in the first model, breadwinning for men and home-making for women are prescribed, but in an 'imperfect' and different way. While women are entitled to derived social rights (e.g., deceased husband's pension entitlement), several social security measures are only partially informed by the 'family wage' principle. As a matter of fact, in this model intergenerational ties and kin solidarity along the life course are expected both within the household and between households.

Social support and obligations between kinship members might even be defined explicitly by law. However, social policies are often not explicitly designed to achieve the desired family/kinship solidarity model, which is somewhat taken for granted. As a result, we find several laws and social policies that only indirectly or implicitly support the family/kinship solidarity model. Social solidarity and family dependencies, even among adult citizens, are institutionally recognized and supported as valuable. For example, survivors' pensions might be granted to the economically dependent relatives (i.e. parents, siblings) of deceased pensioners. Parents' entitlement to the survivors' pensions of their offspring might be seen as a reward for having provided financial and caring support for the dependent worker while he/she was still young. Likewise, tax allowances might be granted to taxpayers for providing care and income for their elderly parents. This system of entitlement can be seen as encouraging intergenerational solidarity. In addition to positive policies, sometimes the absence or the minimal development of certain types of social policies and/or publicly funded services supports a certain model of the family and the accompanying division of responsibilities between the state and the family. If the generally desirable model of family behaviour is one in which caring work is done primarily by the family (women), we would not expect to see well-developed social services for children and the elderly. To conclude, this model suggests that social policy and law related to the family is founded on the following assumption: the family/kinship network is the main source of social protection in society and state intervention in this domain is legitimate only in the case of family/kinship failure. As a result, in the third model a high level of solidarity within the household between kin members should be found.

The characteristics of the three models, as illustrated in Table 2, did not exist in a pure form in any country. In addition, one has to take into account

Table 2
FAMILY MODELS: HOW HAVE LAWS AND SOCIAL POLICY DEFINED THE FAMILY

	Male breadwinner model	**Dual-earner family model**	**Family/kinship solidarity model**
Assumptions about family functioning	*Nuclear family*	*Individual*	*Kin*
Assumptions about division of labour	Gender division of labour within the nuclear family	Dual-income family model	Family/kinship as a 'compositor' of different resources and incomes
Labour market and gender relations	Full male employment	Full employment	Male employment shortage and subsidiary female employment
Family relationships, marriage, and family values	Diffusion of single-person households among the young. Low importance of kinship networks	High diffusion of pluralistic household forms	Strength of family ties and kinship networks (e.g. high frequency of contacts between family/kinship members living in different households, generations live longer together).
	Medium level of marriage institutionalization (medium rate of cohabitation and divorce, medium rate of births outside wedlock)	Diffusion of different family formation forms, low importance of marriage	High institutionalization of marriage (low rate of cohabitation and divorce, few births outside wedlock)

Table 2 (Continued)

	Male breadwinner model	**Dual-earner family model**	**Family/kinship solidarity model**
The family as defined in laws and social policy	*Emphasis: nuclear family*	*Emphasis: individual*	*Emphasis: kin*
Legal obligations between 'adults'	Legal obligations between family members (i.e. father and children)	No	Legal family obligations are extended between kin (i.e. siblings, parents-in-law)
Insurance benefits related to the family	Survivors' pension paid to widows and dependent children	No	Survivors' pension paid to widows, dependent children. But in some cases, also to parents, siblings, grandchildren and others
Tax policies	Family is the unit of assessment. 'Quotient familiar' or income-splitting	The individual is the unit of assessment. Tax credits for dependent children	The individual is the unit of assessment. Tax credits for dependent children, wife and kin
Family benefits	High family benefits for wife and children	Basic universal family benefits	Very low family benefits and for all dependent family and kin members
Child-care services	Not well-developed	Well-developed	Underdeveloped, except for the one framed within the educational system

that they present many limitations – they are not differentiated from a social point of view, and from a temporal point of view they are described in a static manner. Clearly, from a diachronic perspective some countries have moved from one model to another according to different historical phases of welfare state development and in relation to family changes. For instance, Scandinavian countries have passed from a *male breadwinner* family model prevailing up to the 1950s and 1960s, to a widespread *dual-earner* model since the 1970s. Despite the limitations of the typology, it constitutes a useful analytical tool for guiding the empirical research on the two historical cases.

Starting from the illustrated theoretical perspective and keeping in mind the typology presented in table 2, the following chapters reconstruct how laws and social policies have defined the family in Mediterranean welfare states. In particular, in reconstructing how public and private obligations and responsibilities have been legally defined with regard to the need of economically dependent individuals, I argue that it is necessary to focus on the issue of the allocation of family responsibilities and the question of 'familial dependence'. Before going in depth into the analysis of the two historical cases – Italy and Spain – the next chapter will illustrate the importance of focusing on this issue of family responsibilities and familial dependence through an historical overview of the emergence of state policies towards families across European countries.

NOTES

1. For an analysis of the historical typology of the relationship between the state and the family, see Balbo (1987).
2. See Esping-Andersen (1990).
3. By 'feminist approach' I refer to analyses that take into account gender relations as both cause and effect of various social, political, economic and cultural processes and institutions.
4. For a criticism of feminist welfare state theory, which only describes 'women's oppression' without developing a framework for gender analysis, see Daly (1994).
5. For an analysis of Scandinavian literature on the welfare state and the feminist approach, see Leira (1993).
6. For an overview of the Italian literature on this approach, see the symposium Famiglie e Politiche Sociali, in *Inchiesta* 65 (1984). In particular see the articles by Trifiletti (1984), and Balbo (1984).
7. An interesting interpretation of Italian sociological and feminist approaches to the welfare state is offered by Bimbi (1993).
8. According to Orloff (1993: 305), the *power resources* school of analysis 'has demonstrated that *politics matters*, in contrast to those who contend that social policy simply reflects the systemic *needs* of capitalist, industrialised societies'.
9. According to Pierson (1993), the concept of *policy feedback* is of particular importance; the idea is that previous policy choices as institutionalized in particular arrangements affect continuing policy developments whereby, as the author puts it, 'effect becomes cause'.

10. Ferrera's argument is that the first decision about coverage, that is, to whom to give protection, influenced the organizational form and has maintained a strong persistence over time (Ferrera, 1993: 77).
11. See, for instance, Lewis (1997).
12. See Orloff (1993), and O'Connor *et al.* (1999).
13. For a review on the current state of studies on the welfare state and gender see Orloff (1996).
14. According to Koven and Michel (1993) 'maternalism' can be defined as a set of ideologies and discourses which emphasize the role of woman as mother and give to that role (care, nurturance, and so on) societal value. Other historians have preferred a restricted definition that contrasts 'maternalism' with 'feminism'. Finally, historians such as Skocpol (1992) distinguish between 'maternalist' and 'paternalist' welfare states. According to Skocpol, a 'maternalist' welfare state should work for the 'good' of women and their children.
15. Although many, frequently loaded terms have been used to refer to Mediterranean welfare state countries (i.e. 'late-comers'), here I refer to them only by their geographical labels.
16. Note that France is not included in this cluster of countries.

Chapter 2

STATE POLICIES TOWARDS THE FAMILY

Introducing the family into an analysis of the evolution of social policies shows that in all European countries the issue of family responsibilities became a central question in the discussions which preceded the approval of the main schemes of social insurance (Pedersen, 1993; Saraceno, 2000). This chapter focuses on the emergence of state policies towards families in most European countries.

From a theoretical point of view, to deal with state policies towards the family means looking at state policies regarding familial responsibilities (who, within the family, is defined as being responsible for whom) and the corresponding question of 'familial dependence' (who, within the family, is defined as being dependent on whom). In the previous chapter it was argued that this is a crucial question because it illuminates how family (private) obligations, and the related state (public) obligations (i.e. the acknowledgement of the public responsibilities for dependent family relationships) have been framed, defined and supported in specific areas of social policies and/or in legal norms. In particular, there are some questions related to private and public obligations and responsibilities that will underlie the whole analysis of the two historical cases illustrated in this book: when and how did the state opt to intervene in family matters? and which forms of family responsibilities have been encouraged by state policies over time?

This chapter aims to achieve two, more limited aims. The first is to illustrate when (from a temporal perspective), by which social circumstances and actors, and how (by which policy instruments) the governments of the various European countries decided to assume part of the family responsibilities and family dependencies as part of public responsibilities. To this end a historical overview of the main phases of the emergence of state policies towards family responsibilities across European countries follows. The introduction of family policy into different traditions and typologies in the second section will also help readers to place and to understand the different traditions of state policies towards the family.

The second aim of this chapter is, in fact, to place Mediterranean countries, specifically Italy and Spain, within the wider framework of comparative welfare state and family policy developments. To this end, in the third section of this chapter differences between Italy and Spain, on one hand, and other historical cases in regard to state policies towards the family on the other, will be highlighted and the specificity of the southern European welfare states stressed.

The Emergence of State Policies: A Historical Escursus

In most European countries the origins and development of state policies towards family responsibilities (and towards the issue of family dependencies) can be dated from the period between the turn of the twentieth century and the Second World War. This epoch was characterized in most European countries by the emergence of two issues related to family life. First was the issue of poverty and the associated high levels of child and infant mortality. In this period salaries were low, unemployment was widespread, and working and living conditions were poor. Most political leaders connected these high rates of child mortality not just with poverty, but also with unhealthy working conditions and a lack of knowledge about the care of children among mothers. Population decline was the second important family issue that attracted government concern before the Second World War. The period from 1870 to 1939 was marked by major changes in family size. Fertility, which averaged four or five children per woman in the early 1870s (i.e. England 4.94; Sweden 4.11; France 3.42), declined to levels close to two children per woman in 1930 (i.e. England 1.95; Sweden 1.96; France 2.27). While population and family issues attracted little attention prior to the 1920s (with the exception of France), the situation was rapidly altered as a result of the above-mentioned major demographic and economic changes. The decline of fertility raised alarms in several circles and was a subject of major political concern in many countries. In Italy and Spain fertility declines were observed from the early nineteenth century, but it was not until the 1920s and 1930s that the decline escalated and become pervasive in Italy. In Spain the decline became most dramatic in the 1940s (Delgado Pérez and Livi-Bacci, 1992).

To restate, concern about the effects of poverty and fears about population decline were the main forces driving governments to initiate support for families and acknowledge family responsibilities. Three sectors were the target of government intervention towards families: health and welfare services (social assistance) for mothers and children; financial support for families; and support for working mothers. Measures introduced during the first half of the twentieth century under each of these headings constituted the first elements of 'family policy' (Gauthier, 1996).

The first state policies aimed at the issue of family responsibilities had already been initiated in most European countries by the turn of the century. These first measures were not specifically targeted at families, but rather at mothers and children. They consisted of maternity policies and medical assistance programmes for mothers, and child welfare. Since high levels of infant and child mortality were seen as a consequence of the bad working conditions of mothers, several measures were targeted at working mothers in order to improve their conditions before and after childbirth.

From the end of the nineteenth to the first decades of the twentieth century most European countries adopted legislation allowing women or making it compulsory for them to take leave from work before and after confinement (Ballestrero, 1979). International conferences (Berlin, 1890; Bern, 1906) inspired the first discussions, projects, and debates concerning the introduction of compulsory leave throughout Europe. As a result, many countries approved the first legislation on women's labour. For instance, in 1879 Switzerland approved a compulsory eight weeks leave for childbirth. In 1883 Germany authorized the first maternity insurance programme for mutual aid associations. In 1911 this law became compulsory for all mothers employed in industry (Cuesta Bustillo, 1988; Gauthier, 1996).

Compulsory leave was the first step. Next followed the introduction of cash benefits to be paid during maternity leave. Germany was the first country to introduce maternity leave benefit in 1883, followed in 1894 by Belgium and then in 1910 by Italy. In addition, most European countries coupled maternity policies with assistance programmes for child welfare. According to Bock and Thane (1991), between the end of the last century and the 1950s, mother and child assistance programmes provided a wide range of services: hospital services to mothers and their children under a given age, maternity hospitals, 'home helps', food for expectant and nursing mothers and for children under age 5, crèches and day nurseries, and homes for the children of widowed and deserted mothers and for illegitimate children. In some cases the legislation provided only for poor families, while in others it covered all mothers with young children.

As far as the decline in population is concerned, it has to be said that up to the first decades of the twentieth century very few countries assumed explicit measures against the fertility decline. One exception was as already mentioned, France, where fertility had already declined to three children per woman by 1870. France and Sweden were the first countries to seriously address the issue of low fertility and family decline. Concern about low fertility and pro-natalist sentiments became dominant some decades later in Germany, and, as we will see later, in Italy and Spain.

The second phase of welfare state development, expansion, occurred in many European countries between the two world wars. Cash benefit schemes for families were introduced during this period. Cash benefits were first targeted at 'deserving' groups, and were dispensed as widows' allowances, orphans' allowances, and later, family allowances. Britain first introduced allowances in 1914 for the widows and orphans of soldiers. In 1925 these allowances were transformed into a widows' and orphans' pension as part of the Contributory Pensions Act. Other employment-based welfare states also experienced this trajectory of development: survivors of insured workers became entitled to survivors' pensions. By the end of the 1920s many European countries had already introduced a general scheme of family allowances (Gauthier, 1996; Wennemo, 1992). France introduced a sort of 'family wage' *(sursalaire familial)* in the first decades of the twentieth century, when employers set up 'equalization funds' (*caisse de compensation*) to control wages. In 1932 the first *allocations familiales* were introduced, but were not universally extended to wage-earners and the self-employed until 1938. Only families with more than two children were eligible for this benefit (Villac, 1993).[1] In Germany *Kinderbeihilfen* was introduced in 1935 and assumed a pro-natalist orientation in 1939, when it was decided to grant it only for third and subsequent children. In National Socialist Germany, family allowances originated in a context dominated by strong eugenic objectives (i.e. improving the 'quality' of the race by preventing the 'unfit' and 'inferior' from being born) – a dimension which would eventually be widely condemned by the international community. As I shall show in the following chapters, family allowances (*assegni familiari*) were introduced in 1936 in Italy and in 1938 in Spain (*subsidio familiar*).

During this same period of expansion, most countries introduced income tax allowances for heads of family, usually husbands and fathers, for their dependent wives and children. These were introduced in Britain in 1911, France in 1917, Italy in 1926 and Germany in 1934 and 1939 (Bock and Thane, 1991). Finally, most countries that had already approved the first insurance scheme, besides widows' and orphans' allowances, began to extend coverage to the dependent family members of insured wage workers. This was especially the case for health insurance.[2] With the exception of the pioneer, Norway (1909), health benefits (hospital and medical assistance) were generally extended to family dependent members between 1930 and 1945.

Differences in Family Policy Traditions

Cross-national comparative analyses of family policies have increased enormously in the last few decades (Gauthier, 1999). However, it is necessary to assert that 'family policy' is neither a clear field of study, nor does it constitute a well-defined component of welfare state research, in contrast to other elements such as pension, sickness insurance and unemployment. The heterogeneity of welfare programmes included under the heading of 'family policy' is the first problem that emerges when reviewing family policy literature. What constitutes family policy? What are its main components? On the basis of which criteria is it possible to argue that a country has a family policy system? The incertitude in finding a common definition of family policy is further complicated by the fact that family policy is characterized by strong variability over time and space. Major variations can be observed in the orientations, the nature and the objectives of family policy. In some countries they have been connected to demographic issues, in others to child poverty. They may have been introduced to support the male breadwinner ideology or gender equality.

These are some of the main questions addressed in comparative family policy literature and they remain partly unsolved. This is because a large range of measures adopted by governments, ranging from legislation on domestic violence to cash benefits for poor families, and legislation for working mothers, are usually included in a broad definition of family policy.

Family policy has to be imagined, in fact, as a 'wide umbrella' of policies, an umbrella that encompasses policies which, implicitly or explicitly, 'aim at enhancing family stability and well-being' (Wilensky *et al.*, 1985: 56). Kamerman and Kahn (1978, 1990) argue that the term 'family policy' may be used to describe what government does to and for children and their families. In particular, according to them, the term has been used to include all those public policies that are aimed at enhancing the situation (the well-being) of families with children and all those policies that have clear consequences for families, even if that impact may not have been intended.

The difficulties in finding a common and acceptable definition of family policy capable of crossing national borders, and the changing meaning of that definition within a single country over time, have not hampered the production of a large amount of comparative family policy research, distinct precisely by the type of branch of social policy studied. In addition, the fact that in some countries, for instance Italy, the absence of an explicit family policy and the reluctance to use this label to refer to welfare programmes which strongly affect family well-being have obviously resulted in a lack of interest and scarce research on this field up until very recently.

In terms of comparative family policy studies, many different typologies of the family policy model have been proposed and these depend on several factors. For instance, the way in which the problems surrounding definition have been solved, the different indicators used, the length of the historical period under analysis, and, of course, the approach adopted are all factors which contribute to several different typologies of family policy. In terms of family policy outcomes, Kamerman and Kahn (1978), in a study which adopts an institutional approach and which is now a classic in the field, distinguish three different models: an explicit and comprehensive family policy model, found in countries such as Sweden and France; a sectoral family policy model developed in countries such as Austria and Germany; and, finally, an implicit and reluctant family policy model, which has characterized the United Kingdom and the United States. On the basis of the explicit/implicit dimension, it can be argued that democratic Italy and Spain also belong to this latter cluster of countries. Gauthier (1996), who also adopts an institutional analysis of family policies from a historical perspective, proposes distinguishing between four typologies of the family model: pro-natalist family policy in France; sex-equality-oriented family policies in Sweden; traditional family policy in Germany; and non-interventionist family policy models in the United Kingdom and the United States. Gauthier argues that in most countries the decline of fertility and increasing female labour participation have been the most important factors in giving an impulse to the development of stronger and/or coherent family policy programmes.

There are few countries that have developed an explicit, comprehensive, coherent and pro-natalist system of family policy, the notable exceptions being France, Belgium and Luxembourg (Kaufmann, 1996). These countries have been, on occasion, implicitly or explicitly been taken as reference points for understanding what family policy is. Indeed, France has been, together with Belgium, a pioneer in the development of family policy. Over the years successive French governments have implemented an impressive array of measures designed to encourage family building and to create an environment in which the welfare of children is recognized and overtly supported by the state. As has been previously mentioned, France was (together with Sweden) one of the first countries in Europe to address the fall in birth rates, taking action as early as the second half of the nineteenth century. Throughout the twentieth century this nationalist and pro-natalist concern remains one of the main causes for explaining the importance attributed to demographic objectives in the French welfare state. In particular, according to Villac (1993) three waves of thought, from the defeat of 1870 up to the Second World War, shaped the political discourses on which the future developments of family policy would rest:

- The *individualist* approach, which was mainly inspired by revolutionary ideas and supported personal freedom and individual choices.
- The *pro-natalist* approach, which was fundamentally concerned with increasing fertility.
- The *familialist* approach, which gave priority to the ethical principles of the family.

In addition to these three waves of thought, some authors note other important historical factors that have shaped the French family benefit system (Villac, 1993; Friot, 1997). According to these authors the foundation of the current and generous family allowance system can be found in the initiatives of a number of employers, most belonging to the social Christian movements, who created the first family allowance funds in the first decades of the twentieth century. Employers' concerns were not only with pro-natalism and/or familialism, but also social integration (i.e. working-class integration), at a time when the incipient industrialization and urbanization processes were creating high rates of poverty. Later, in the 1930s the expansion of the family allowance system in France, and the generalization of *sursalaire familial*, became the basis of some agreements between social Catholics and employers. The former supported family benefits because they served familialist and pro-natalist ideas. For the employers, family allowances became a good policy instrument to control wage increases while meeting the demands of the working class. In the previous chapter I mentioned that in recent decades the French welfare state model has come to be seen as one in which the state recognizes and promotes woman's position as both worker and mother as well as an illustrative example of the 'parental model' of the welfare state.

As far as the non-French-speaking European countries are concerned, to find an agreement among authors over their development and family policy traditions is not so easy, because of the lack of coherence and continuity in family policy models in most of these countries (Gauthier, 1999). For some of them, for instance the Scandinavian ones (Sweden is always taken as an example), it might be argued that they have developed not so much an explicit system of family policy focused on the family unit, but rather a system of social policies focused on the needs of children and on gender equality. In other words, it is not so much their continuity of policies and the explicit dimension of their family policies, but, since the 1970s, their objective of gender equality, their support of the 'dual-earner' family model and individual autonomy that is more visible in the case of Scandinavian countries. This objective has been achieved mainly through the development of a system of child care and of family-friendly and father-sensitive policies (long parental leave) that have reduced gender inequalities, intergenerational obligations and encouraged the autonomy of the young.

Germany has experienced a different family policy tradition, both with regard to the Swedish case and to the French one. The history of family policy in Germany originated during the Nazi regime, with objectives not only to increase the size of the German population, but also of racial selection. In post-war Germany, family policy was not, in any circumstances, to be confused with population policy. Despite the fact that for the German

Federal Republic case one might argue the development of an explicit family policy system (see Kaufmann, 1996), its social protection system has not been as generous towards families with children as the French or Belgian ones. It is far from the French case also because of its traditionalism as far as the division of gender roles within the family and in the labour market, and for the target being the 'complete' and legitimate family. For instance, German fiscal policy has underpinned the traditional model of the married couple morally bounded together for life. The wife is rewarded for being a housewife and, if employed, for being on a very low income. The tax system favours married couples, whether or not they have children (Hantrais, 1994). In addition, and in contrast to the French family policy model, the German model has been marked by incertitude in its planning and by discontinuity (Kaufmann, 1996).

The historical case of the United Kingdom has often been mentioned as one of the best illustrative examples of a strong non-interventionist policy of government in family matters. Unlike France, governments in Britain conceive the family first of all as a private matter and do not aim to help women to combine professional and family life. The number of nurseries and pre-school places is amongst the lowest in Europe (cf. appendix 8). However, the United Kingdom has been capable of developing a wide range of social policies against poverty, and so it has also provided a minimum income to families with children (Kaufmann, 1996).

Shedding light on differences in family policy traditions might have contributed to revising and discussing the theoretical framework of welfare state research, which so far has not systematically included in the comparative framework neither the issue of family arrangements and changes nor the field of family policies. From a family policy perspective, for instance, the French welfare state model poorly fits the conservative and corporatist regime type. It is particularly hard to group France together with Germany, Austria, Italy and Belgium when analysis focuses, as shown by Pedersen (1993), on gender and intergenerational relations supported by social policies.

The most important features that set the French case apart from other conservative and occupational models of welfare states are: (1) its concern with families with children and more specifically with the importance attributed to demographic objectives (a focus which is shared only by Belgium); (2) a model of employment policy designed to make childbearing compatible with mothers' employment (a focus which is not common to other conservative–corporatist countries, but shared with Scandinavian countries).

More recently other family policy studies have focused, using different approaches and selecting different welfare programmes, on comparisons of public support to families (see Gauthier, 1999). For instance, Bradshaw and

Ditch (1993), focusing on the variability in governmental support and comparing different types of families, have found a high level of government welfare effort in France, Luxembourg, Norway and Belgium, a medium level of effort in Denmark, the United Kingdom, Germany, Australia and the Netherlands, and low level of effort in southern European countries, Ireland and the United States.

Despite the heterogeneity of issues tackled, and of welfare programmes and legislation included under the heading of family policy studies, the increasing interest in family policy studies might help to introduce the family fully in the analysis of social policies and eventually to reframe comparative welfare state research.

The Peculiarity of the Southern European Countries

According to Giner (1985: 310), Greece, Italy, Portugal and Spain have 'an unmistakable commonality and distinctiveness within the larger framework of European society'. These countries display a number of common traits with respect to their historical evolution, modes of political domination, the form and time of economic development and their system of class relations. All four countries have experienced 'despotic regimes' during the twentieth century, all have made the transition to democracy, and all have had a socialist government for some or a large part of the post-transition period.[3]

Socio-economic Developments in Southern Europe

The most important difference between the continental and northern European countries and those of southern Europe (with the exceptions of northern Italy and Catalonia and the Basque country in Spain) is that the latter experienced a relatively late period of industrialization. The main consequences of this lag in development for southern European countries are evident in: (1) the comparatively lower percentage of the total population involved in the labour force; (2) the slower decline in the number of agricultural workers without prior peaks of industrial employment; (3) a rapid growth and a rapid change from a mostly agrarian society towards a service-oriented society; (4) higher levels of emigration where southern European countries supply labour to the Continent; and (5) persistently higher total unemployment rates with respect to continental and northern countries (Sapelli, 1995; Fua, 1980).

The singularity of this socio-economic development also underlines some very important aspects of the socio-economic structures of the agricultural sectors of these countries. First of all, there have been huge territorial imbalances. In particular, in Italy the north–south cleavage and the economic backwardness of many regions still call for continuous and active support from central government. In Spain the territorial imbalances have been between the early industrialization of Catalonia and the Basque country and the backwardness of the other regions. Second, property/land ownership has been extremely unbalanced. For instance, while land ownership in the north of Spain has been characterized by a stable structure of smallholdings, property in the south has been distributed among large estates (*latifundios*). A similar heterogeneity in the structure of landownership is still found in Italy between large and medium-sized capitalistic agrarian concerns and small family-run farms (Fabiani, 1979; Pérez Yruela, 1990).

As a result of these southern peculiarities, comparisons between southern Europe and continental Europe reveal a historical variation in the relationship between the primary and the secondary sectors during the development of the industrial economies and modern societies.

Southern European Welfare States and State Policies Towards Families

The history of European welfare states, and more recently, the history of southern European welfare states has already been analysed by many scholars. As the first chapter has shown, academic debates on the welfare state have now started to systematically include the Mediterranean countries within their scope of observation. There, the existing interpretations of the historical evolution of southern European welfare states have already been outlined. The purpose of this book is to present a different story, that is, to tell when and to what extent the Italian and Spanish welfare states began to include the issue of family responsibilities within their systems of social provision.

In Italy and Spain neither an explicit family policy nor a high level of state support to families with children exists. This places these two southern European countries in strong contrast to the French case, which displays an explicit, visible, coherent and pro-natalist system of family policy, a generous level of child benefit and child-oriented and parent-friendly social policy. Despite sharing the Mediterranean pattern of delayed industrialization and Catholic identity, France is very far from the low development of social policy related to families that is seen in democratic Spain and Italy.

This differentiation also shows that the historical evolution of state policies towards family responsibilities, from a gender and intergenerational perspective, is crucial for understanding the general evolution of the welfare state. It may, in fact, help to explain why countries categorized as belonging to the same 'family' or welfare state regime, such as France, Austria, Germany and Italy, developed along such different lines with respect to gender, generation and kin dimensions. The Italian and Spanish cases, in fact, seem to differ from most of the other countries specifically in terms of a different development of the welfare state with regard to the allocation of responsibilities within intergenerational and kin relationships. Specifically, one of the most 'peculiar' features of functioning of southern European countries is how unevenly the reproduction and caring tasks have been performed by families, in contrast for instance to the French case, in which the state has strongly supported families with children. Similar to the United Kingdom case, the attitude of the state towards the family in democratic Italy and Spain has been that of non-interventionism, but unlike the United Kingdom case, a high degree of family burden and obligations in southern countries has been allocated by laws and social policy not only to the nuclear family but also to kin. As Table 2 shows, social solidarity and family/kin dependencies, even amongst adult citizens, have been codified in legislation and encouraged and supported by southern welfare states. This is their most distinctive characteristic.

However, in reading the following chapters, it is useful to keep in mind some other features of southern European countries that are particularly relevant for understanding the distinctive origins and development of their social policies toward families: (1) the late timing and qualitatively different form of industrialization and modernization, which have resulted in different labour market structures and territorial imbalances; (2) the central role of the Church and its social doctrine, its emphasis on the principle of 'subsidiarity' and its monopoly in running public poor reliefs and services for families; (3) the 'despotic regime' which Italy, Spain, but also Greece and Portugal have experienced; (4) the development of welfare states along occupational lines, characterized by a high degree of institutional fragmentation, a low penetration of the welfare state and low degree of stateness, and an unbalanced distribution of risks; and finally (5) the presence of common cultural and value patterns within the southern European countries, especially concerning their distinctive system of family and kinship solidarity.

NOTES

1. Among the huge literature on the history of family policy in France, for a general overview see Villac (1993). For an interpretation of the issue of family allowances as a form of wage negotiations between employers and workers, see Friot (1997).
2. Health insurance in occupational-based welfare states has consisted of two different types of benefits: cash benefits and health benefits. Cash benefits, i.e. daily sickness indemnity, or maternity indemnity, have been provided only for insured workers. Health benefits, including hospital, pharmaceutical and medical assistance, have been provided both for the insured worker and his/her dependants and for pensioners.
3. However, this argument only partially applies to the Italian case. It is worth remembering that in Italy the Socialist Party (PSI) has never been one of the two largest political forces. It has always shared power in coalition with the Christian Democrats (DC).

PART TWO

ORIGINS OF STATE POLICIES TOWARDS THE FAMILY IN THE AUTHORITARIAN PERIOD

INTRODUCTION: DEFINING FASCIST-ERA FAMILY MODELS

The second part of this book illustrates the evolution of state policies toward family responsibilities in Italy and Spain, from their origins to the fall of the fascist dictatorships (1922–43, 1939–75 respectively). I wish to identify the first steps taken by the Italian and Spanish states to support families, or rather, to define which particular forms of family responsibilities and of dependent family relationships were worthy of public support. When and how did the state opt to intervene in family matters in the Italian and Spanish cases? Which forms of family responsibilities and of 'familial dependence' have been encouraged by state policies? What definition of family models has been incorporated into social policies? These are some of the underlying questions which will be answered through the analysis of the Italian and Spanish historical cases. The third chapter reconstructs how social policies and laws have defined the family in the Italian case, while the fourth chapter deals with the Spanish case.

Analysis of the two historical cases tests the following hypothesis: by the end of the authoritarian regime, Italy and Spain had only partially achieved the version of the male breadwinner family model prevalent in other European welfare states. Instead, since the very first family-related policies, Italy and Spain have tailored their male breadwinner model towards a family/kinship solidarity model in which family and kin interdependencies and intergenerational responsibilities along the life course were expected and were sometimes supported by public provisions.

In reconstructing the origins and development of state policies towards family responsibilities in the Italian and Spanish cases, the structure of the analysis follows a similar framework. First, the socio-economic structures, the legal context of legislation concerning the family, and the general development of the two welfare states are explored. This analysis is important because it provides the socio-economic context and the institutional framework in which the first state policies towards dependent family relationships were embedded.

Both countries underwent relatively late and substantially different processes of industrialization compared to other European countries. In particular, up until the 1950s, both countries had been marked by the prevalence of the agricultural sector, with changes in Spain lagging behind those in Italy. Specifically, the very low rate of female labour force participation in the two countries over a long period of time provides a useful framework for understanding the origins and the institutional development of maternity law throughout the century.

In relation to the legal context of the family, analysis of the two national cases will be developed to look at how the reciprocal legal duties between the spouses, between parents and children, and between kin have been regulated by the main branch of what is called family law. Family law is mainly dispensed in the civil code, although some principles are often mentioned in the Constitution and in fiscal and administrative laws. Given the wide spectrum of relationships covered by family law, the analysis focuses on marriage and divorce law, laws governing parenthood, and kin-maintenance law. The analysis thus illustrates how relations between men and women, across generations, and among kin have been regulated and enforced by civil law and how they changed with the advent of the two dictatorships. In relation to this, it will show that neither the Franco regime nor the Mussolini dictatorship created new legal definitions of the family. Rather, in Italy the fascist civil code (of 1942) revived earlier family law traditions and strengthened authoritarian principles and gender inequalities. Franco's abolishment of the republican laws reinstated the old nineteenth-century civil code, which stressed female subordination to marital authority. The only clear changes in the legal definition of 'family' were in the definition of abortion and marriage. Abortion was defined as a crime against the state.

The family model supported by the two dictatorships was both traditional and 'corporatist'. The family was regarded as a basic cell in the body politic and as an institution organically linked to the state, which one was expected to be a component and instrument of. In this system women were assigned mainly to home-making, discouraged from entering the 'official' labour market, and confined to 'unprotected' jobs. The legal status of women during the two dictatorships was almost totally dependent on their husband/father. By contrast, men were regarded as breadwinners, they were legally defined as the heads of family, and were actively supported as such through social policies. At the same time the Italian and Spanish dictatorships regarded women as important producers of the 'race' and protectors of tradition.

The development of the Italian and Spanish welfare states has been marked by the gradual introduction of social security systems based on a similar 'occupational model' of the welfare state. However, this development occurred first in Italy and then in Spain, the latter being a late-comer compared to other European countries. Most western European welfare states had already established a social security system by 1950. At that time Italy had introduced general schemes which protected workers and their dependants from injury, old age, disability, unemployment and illness. In contrast, in the 1950s Spain did not have a general scheme for unemployment and its health insurance system only provided minimal coverage to low-income dependent workers and their dependent family members.

These variations in general welfare state developments between Italy and Spain might be accounted for by the different level and timing of the industrialization processes, and by the slow decline of agricultural labour forces, Spain ranking last in comparative perspective. But, surely, they must also be related to the advent of civil war (1936–39) in Spain and to the long dictatorship, which delayed the development of a modern welfare state until the 1960s. Finally, in both countries the development of the welfare state has been marked by the ubiquitous influence of the Catholic Church, which often competed with the state for control over social protection and welfare provision. In particular, the Church managed to maintain its hegemony in certain welfare state domains that included social assistance, social services for children, educational policy and family law.

In illustrating the two national cases, the analysis intends to look beyond the history of welfare state development. It reconstructs how social policies and laws have defined the family models that characterized Italy and Spain, from their origins to and throughout their authoritarian periods, and documents how they have changed over time. After having provided a brief picture of the socio-economic and legal contexts, the analysis focuses on early state policies towards family responsibilities, as they have been incorporated into *maternity insurance, provisions for children, social insurance schemes* and *family benefits*. By *family benefits* I refer namely to *family allowances* and *tax allowances*. In doing so I stress: (1) the temporal variation (taking as a reference point the state's policy towards families as well as general welfare state development in other European countries); (2) the qualitative illustration of the types of family responsibilities acknowledged by the state; and (3) the social categories that were progressively covered (industrial workers, agricultural workers and so on). Another purpose of the analysis is to uncover variations that account for differences in the Italian and Spanish outcomes. I expect particularly important variations in the existing associative structures, in the occupational structure, in the strength and active presence of progressive parties and trade unions, and in the role played by the feminist movement and/or the Catholic Church.

Through the illustration of the two national cases, the reader will identify the points of origin of the first family policies, which in both countries has to be found in the period preceding the rise of fascism. The first state policies towards the issue of family responsibilities, in fact, have to be found in maternity insurance laws. These laws were already approved at the beginning of the century, in 1900 in Spain and in 1902 in Italy. Maternity benefits were the next step forward. They were established first in Italy in 1910 and then in Spain in 1923. Maternity benefits were chiefly located within labour legislation and affected only a very limited number of working women. Despite the limited range of coverage, they are important measures because they mark the first public acknowledgement within social policy of the time and wages women forfeit to produce children.

From a comparative perspective the first maternity policy (compulsory leave period for working mothers) was in line with other European countries. However, if these measures are compared to the Italian and Spanish processes of industrialization, then one can argue that they were introduced quite early. This may be due to early industrialization in the northern regions of Italy and Catalonia and the Basque country in Spain. Indeed, in some industries, especially in Italy, women made up a significant part of the workforce by the turn of the century. Women's labour legislation in both countries was highly influenced by international pressure, but the women's and labour movements also played a very important role in Italy. Spain's laggard position with respect to Italy concerning maternity benefits (not maternity leave) might be related to: (1) the later development of social insurance schemes in general; (2) the lower level of female labour force participation in the 'official' labour market; (3) the active role played by the Italian women's and labour movements in the creation of the first maternity insurance funds.

Apart from the earlier introduction of the maternity insurance scheme, the period of real development of policies in relation to family responsibilities in Italy has to be placed between the two world wars. This was also the case with the other European welfare states. These state policies were first targeted at dependent industrial workers. Only later and gradually were insurance schemes extended to the other sectors and/or categories of workers. This was strictly in line with what occurred in other countries, during what we have described as phase 'two' of the welfare state: enlargement. Thus, because Italy developed social insurance schemes earlier than Spain, we also expect it to have developed an earlier system of state policies concerning dependent family members of workers. This was exactly the case. In Italy the development of state policies for dependent family members originated within the framework of the first social security schemes protecting the dependent industrial worker (i.e. old age and unemployment schemes in 1919), immediately prior to the rise of Mussolini.

The Italian dictatorship expanded already existing measures, and then developed, in line with the fascist ideology of the family, the first Italian family policy system. What has to be underlined is that these first measures assisting with family responsibilities were aimed at solving problems created by the onset of the industrialization and urbanization processes. Therefore, these policies were mainly designed to meet the needs of the new worker – adult male industrial workers who lived in urban areas. They applied, however, to a workforce that in reality was dominated by agricultural workers. Moreover, they were initiated at a time when female labour force participation in the 'official' labour market was declining but participation in informal, 'unprotected' jobs was high. In Spain, the issue of family responsibilities and that of 'familial dependence' was widely raised with the advent of Franco. The issue developed within the ideological framework of the dictatorship, and was clearly coupled with the establishment of the first social insurance measures (i.e. health insurance in 1942 extended to family dependent members; survivors' pensions in 1955).

The introduction of inclusive systems of family policies in both fascist Italy and Spain marked a clear break with past governments' treatments of family responsibilities. The family model of the two dictatorships was inspired by Catholic morality and marked by patriarchal principles. Finally, it was characterized by a rural-oriented rhetoric. Because social policies were initially targeted at industrial dependent workers, they excluded the majority of workers, who were employed in the agricultural sectors and those, mainly women, who were occupied in 'unprotected' jobs.

A common characteristic, which was more generally a feature of most welfare states inspired by the 'occupational model', was that they were heavy on transfers (cash benefits) but light on services. In this regard, it has to be kept in mind that in these two countries the Church has managed and run service and assistance programmes for a long time. The two regimes did not hinder, and when they did only to a limited extent, the hegemony of the Church in the domains of family law, 'poor' relief, services for children and education policy.

Family policy developed in Italy and Spain in relation to both the ideological and political economies of the two regimes. In line with the fascist ideology of protecting motherhood, the government instituted public 'agencies' for supporting 'poor' mothers and generating population increases during this period. The most important benefits introduced were Italy's *assegni familiari* (1936) and Spain's *subsidio familiar* (1938) and *plus de carga familiares* (1945). These benefits were introduced first as a means of combatting poverty, but later promoted the regime's more general pro-natalist policy. Family allowances were soon coupled with other cash benefits, which also aimed at increasing the population and supporting the male breadwinner family model. Family benefits were, in fact, paid out to the head of the family (the husband/father). Women could be eligible for family allowances only if they were widows, single mothers, or married to men unable to work.

Family benefits during the fascist regimes, as in many other European countries at that time, aimed also at shaping gender relationships within the family, promoting fatherhood through income support, encouraging women's reproductive capacities, and restricting women's role to the household. Family allowances in Italy were stripped of their pro-natalist dimensions in 1944 with the demise of the dictatorship. This was not the case in Spain, where change only came in 1966 with economic development and liberalization. In 1945 Spain introduced the *plus de cargas familiares,* within the framework of its wage policy system, which supplemented the breadwinner's wage. The *plus de cargas familiares* best expressed the purpose of the regime to create a real 'family wage' by which a male breadwinner could support his dependants without his wife having to work outside the household. As discussed above, agricultural workers were not entitled to this benefit. Also most of the *other family benefits* were paid to the 'head of the family' (tax deductions, marriage allowances, birth allowances) and thus designed to strengthen the male breadwinner role and the homemaking role of women. *Health insurance* and *widows' and orphans' pensions* also supported the male breadwinner family model, insofar as they reinforced wives' and children's dependence on the husband/father. They were, in fact, conceived first of all as rights for insured male workers, and designed to grant women's and children's 'derived social rights'.

Finally, I want to emphasize what the reader will recognize as a specific feature of the Italian and Spanish cases: the extended concept of dependent family relationships. Most of the state policies enacted during the two dictatorships were not targeted only at family responsibilities for an economically dependent wife and children, but also at other dependent family members. In Italy, from 1940 onwards the beneficiaries of family allowances were extended well beyond the nuclear family to siblings, parents and parents-in-law with no or low income. The fascist civil code of 1942 (*codice* Rocco) maintained the legacy of the early kinship obligations. In Spain the *subsidio familiar* was given to the insured worker for children, grandchildren and siblings, with certain restrictions up to the 1966 reform. The *plus de carga familial* covered children and economically dependent parents and siblings. Spain also extended health insurance beyond the nuclear family, covering economically dependent siblings and ascendants (*ascendientes*).

Chapter 3

THE ITALIAN CASE

This chapter provides a general perspective on the Italian welfare state, focusing on the origins and the development of state policies concerning family obligations towards dependent family members. An overview of the socio-economic development of the country is provided first, in order to contextualize the policies of interest. Second, the legal context of legislation concerning the family is presented. The main social policy concerning the family is analysed for two distinct historical periods. The first or 'take-off' period (1902–21) covers the first women's labour law to the rise of Mussolini. The second or 'fascist' period (1922–43) provides an overview of the system of family policy during the regime. As will be seen in the following sections, the fascist regime showed deep concern about population and family issues and manifested a strong position in the regulation of family matters.

Socio-Economic Context

In relation to the long-term development of socio-economic structures, Italy and the other Mediterranean countries have been analysed in two different ways. First, for those who regard industrialization and modernization as step-by-step processes, these countries are seen as 'late-comers' compared to the 'classic' development of other industrialized countries. From the second perspective, they are regarded as countries standing in a half-way position, a position resulting from their economically and politically peripheral (or 'semi-peripheral' in Wallerstein's view) status in the world of capitalist division of labour *vis-à-vis* the 'core' economies.[1] According to this second interpretation, we should speak not of 'late-comer' countries but of a 'peculiar path' of development deriving from a semi-peripheral position in the world economy (Wallerstein, 1979). This position tends to give priority to the specific characteristics of national labour market formation as an effect of the semi-peripheral position in the division of the world's labour market, rather than as a function of later development.[2]

Borrowing this second interpretation, it can be said that since the beginning of industrialization the Italian labour market functioned as a sort of response to demands coming from the most advanced foreign markets. According to Paci (1981), after the 'take-off' years of industrialization the Italian labour market was characterized by several features that would remain stable over the next hundred years. Following Paci's analysis, the main characteristics of the Italian labour market were and still are

- specialization in industry that mainly produces consumer goods for the most advanced markets, which means having a semi-peripheral position in the international division of labour
- the development of a labour force characterized by a structural employment shortage, a precarious working class, and the diffusion of small family controlled enterprises, typical of 'light' manufacturing industry
- a strongly uneven territorial development of the country, characterized by strong migration of the labour force within the country and abroad
- the development of industries in an agricultural setting.

Occupational Structures

In order to provide a very general picture of the long-term developmental trends of the Italian labour force, this section presents basic data on the development of labour force participation by sector of activity (agriculture, industry and services) and female labour-force participation from the late nineteenth century to the 1980s. Before starting to analyse the Italian statistical data, it is important to note that population as well as industrial censuses from Italy suffer from many shortcomings, especially when a diachronic view is superimposed upon them. Monthly surveys on employment started in Italy only at the beginning of the 1950s; valid historical data from before this date are scarce.

The century between 1881 and 1981 has been characterized by a reduction in the Italian labour force, with an increase in the size of the non-active population. This trend began to show some signs of inversion at the start of the 1980s, mainly due to the higher rate of female participation (De Rosa, 1980; Zamagni, 1994).

As can be seen from Figure 1, the most impressive change during the last century has been the shrinking of the labour force in agriculture from 6–23 per cent (in 1881) of the active population to 11 per cent (in 1981). It is more difficult to comment upon trends in the industrial sector because of discrepancies among the various sources. According to the population censuses, participation in the manufacturing sector declined from 18.6 per

Figure 1
LABOUR FORCE BY SECTOR, ITALY, 1881–1981

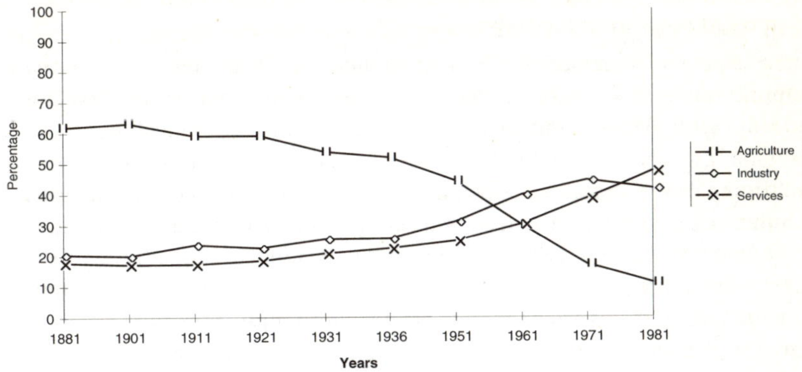

Sources: Based on figures supplied by Zamagni (1994), table A.1, p. 231. Data are calculated on the basis of population censuses, with adjusted figures elaborated by the author. Industry includes: mining, manufacturing, construction, electricity, gas, water.

cent in 1911 to 17.1 per cent in 1921. This decline was followed by steady increases in industrial employment until 1971, when it reached 44 per cent. It then began to decline, falling to 41.8 per cent in 1981. Industrial censuses demonstrate different trends, reporting a continuous increase in industrial employment from around 13 per cent of the active population in 1911 to 30 per cent in 1981. It is reasonable to conjecture that the discrepancy between the population census and the industrial census may be due more to variations in how different types of work were reported rather than in structural causes. It is likely that the small declines reported in the population censuses from the 1970s are the result of how census-takers counted household work, home-workers (*lavoratrici a domicilio*), unemployment and under-employment (*sottoccupazione*).

The service sector remained steady, at around 17–18 per cent until 1921. The growth of this sector was quite slow until 1951, when it subsequently began to increase more rapidly, reaching 47 per cent, or nearly half of the labour force, in 1981. Despite the scarcity of reliable historical data series, we can compare data on the industrialization of the Italian case with those from other European countries. Data provided by Flora, Kraus *et al.* (1987) on the distribution of the labour force by sector demonstrate that while Italy's industrial sector did not outsize its agricultural sector until 1961, Great Britain, Belgium and Germany made the transition at least 50 to 100 years earlier. Austria's industrial sector began to dominate in 1951, while French industry did not overtake agriculture until 1954. As will be seen in the next chapter, Spain's shift from agriculture to industry happened even later than it did in Italy.

Female Labour Force Participation

When trying to understand why and at what point in time states initiated maternity laws, it is important to take into account trends in the levels and sectors (agricultural, industrial and service) of female labour force participation. The structure of the labour market itself provides an important framework for policy. As Pedersen (1993) suggests, the labour market structures of a country help make particular policies imaginable. Policies do not, however, simply reflect patterns of dependence produced by labour market structures. Such a view is untenable because labour market structures in themselves are in part political creations. Pedersen argues, for example, that the weak position of married women workers in Britain was partly the consequence of state policies that marginalized such workers after the First World War. I argue that a similar phenomenon explains the decline of official female labour force participation during the 1920s and up until the 1930s in Italy.

According to the industrial statistics provided by MAIC (Ministro Agricoltura Industria e Commercio; Ministry of Agriculture, Industry and Trade) and reported by Merli (1972: 112), in 1876 the composition of the industrial workforce in a defined number of industries was 48.6 per cent women, 23.2 per cent children and 28.2 per cent men. In 1903 the composition was 53.7 per cent women, 14.5 per cent children and 31.8 per cent men. Children who left manufacturing jobs under controversial child labour laws were replaced, primarily, by female workers. Although the figures reported refer mainly to textile firms that typically employed female workers, it has to be underlined that females also outnumbered males in the paper and tobacco industries. Most of the female industrial workers were composed of unmarried young women, as was the case in other countries. Therefore, at least until the 1910s, certain areas and sectors (especially textiles) of Italian industrial development depended upon the widespread employment of women. This phenomenon can be explained by the fact that women's wages stood at about half those of men during this period. According to Bettio's study (1988, table 5.4) of several cotton, wool, silk, paper and other textile firms located in northern Italy, women on average earned only 43 per cent of men's pay in 1850. A few years later (in 1883) female relative earnings in the wool industry were 50–60 per cent of male's wages. This pay gap existed at both the top and bottom of the wage scale (Bettio, 1988, table 5.5; see also Meldini, 1975: 70–2).

Since the first decades of the twentieth century most European countries, including Italy, have experienced an inversion of trends in the relationship between female and male employment. Women were gradually expelled from the 'official' labour force through provisions that favoured men's employment. Official labour force is emphasized here because the census figures are problematic indicators of women's work experience. The assumption beneath the census categories was that a clear distinction existed between 'occupied' wage earners and 'unoccupied' housewives, whereas, historians have shown, many working-class women moved in and out of the labour market and/or were occupied in marginalized jobs not represented in census figures. At any rate, the official data indicate deep differences between sectors of activity and between countries, though the general trend towards the expulsion of women from the labour market holds across all countries. For instance, in France women's activity rate as a percentage of the total labour force dropped from 35.7 per cent in 1901 to 34.2 in 1936 (Norvez, 1990). In Sweden women's activity decreased from 28.3 per cent in 1900 to 25.5 in 1925. In England a percentage drop from 26.9 per cent in 1891 to 25.5 per cent in 1921 occurred (Meldini, 1975). The decline of women's activity is particularly evident in the Italian case, where it dropped from 32.4 per cent in 1901 to 29.0 per cent in 1911, to 26.8 per cent in 1921, to 19.0 per cent in 1931, and then went up to 24 per cent in 1936. In absolute numbers, 5.3 million women participated in the labour force in 1901, 5 million in 1921, 3.8 million in 1931, and again 5 million in 1936. The strongest reduction between 1921 and 1936 was in the

Figure 2
FEMALE SHARE OF THE LABOUR FORCE, ITALY, 1901–51

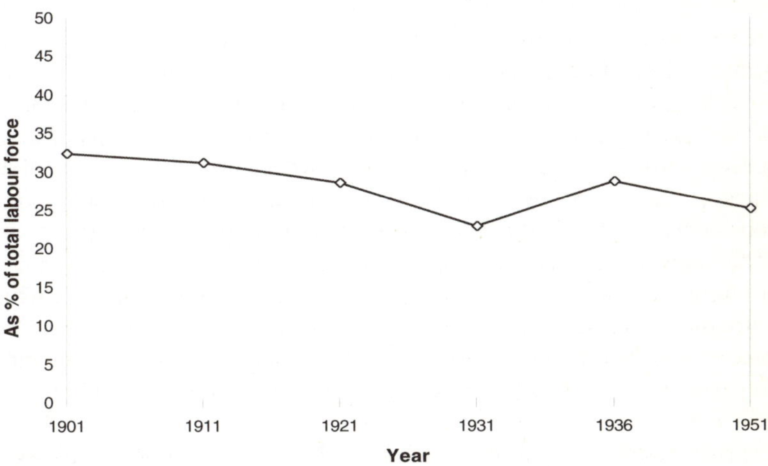

Source: Based on figures supplied by Svimez (1961), pp. 50–1.

agricultural sector, where the number of employed women dropped from 3 million in 1911 to 1.5 million in 1931 (Svimez, 1961: 50–1). But as previously mentioned, the sharp reduction in female labour force participation in the official labour market does not necessarily indicate an actual reduction in women's employment. It is more likely a statistical artefact. For instance, in Italy data on women's employment did not count domestic work – the main source of women's employment.

As can be seen from Figure 2, the female labour force share in Italy in the official labour market has been progressively and steadily declining since 1911, the lowest level being recorded in 1931. Although there was a slight increase from 1931 to 1936, this was followed by a further decline (Federici, 1984). There is more than one reason for the decline in official female activity rates in Italy from 1911 to 1931. The first is the passage from manufacturing to industry. Second might be the growing influence of the labour movement, which opposed the entrance of female labour into male jobs. Third is the economic crisis of the 1920s and 1930s and the introduction of legislation that tended to favour the employment of men in the official labour market (i.e. fixing a threshold level of women's employment in public services). Finally, new legislation protecting women workers made female labour more expensive and therefore less attractive (Ballestrero, 1979 and 1996).

The Legal Context

The Root of Italian Family Law

The first unification of the different laws and customs regulating families occurred in Italy during the invasion of Napoleon. In 1804 the Napoleonic civil code (Code Napoléon) became law in all parts of Italy, except the islands. The Code extended not only to Italy but to all countries occupied by Napoleon's troops. Indeed, its influence went well beyond just these countries. The Code inspired the development of family law in most European civil codes, and has left its imprint up until very recently.

The main characteristic of Napoleonic family law is a clear and unequal distribution of authority between genders and across generations which favours the father/husband. In particular, the family covered by the Code Napoléon was mainly the bourgeois family model where property relations were particularly important. The Code had three main features. First of all, marriage was not regarded as an individual decision but as a family enterprise requiring parental consent. Next, the Code emphasized legitimacy as central to the family and granted maintenance and inheritance rights only to those children born within marriage. Finally, the Code articulated the domination of husbands over wives. Married women were denied the right to take legal action and could not sell, mortgage, or donate their own property without their husbands' permission. Husbands also managed all common property. Husbands and wives had unequal rights in terms of claiming the legitimacy of children.

The Napoleonic code was also distinctive in that it introduced divorce. In the reading of the evolution of family law in Italy, special attention has to be paid to the issue of divorce. This issue, in fact, symbolizes the control of marriage and through it the control of the family. Thus divorce has been one of the main issues at the heart of questionable relationships between the Church and the state. It is exactly because the Code Napoléon introduced civil marriage and divorce that in Italy the Church strongly opposed its introduction. Indeed, in this country divorce was soon forbidden with the Restoration in 1816. During the Restoration, in most Italian states family legislation reverted to being very similar to that of the beginning of the century, based on a fragmented set of rules and customs. At this time the Napoleonic principle of civil marriage was abolished and it again became legal to celebrate marriage according to the canonical law.

In the new united Italian Kingdom (1861) the civil and penal codes were unified for the first time over all Italian territories in Pisanelli's civil code and Zanardelli's penal code (1865). As in many other European countries, the Code Napoléon inspired this first Italian civil code. The new civil code established the secularity of marriage (*laicità del matrimonio*) and the competence of the new state in ruling over the formation and the life of the family. Despite this innovation, Pisanelli's civil code of 1865 was not meant to completely replace or negate Italian social traditions or the state's relationship with the Church. So, despite the fact that civil marriage was recognized as the legal form of the marriage contract, it was never declared superior to those conducted by the Church – a strong contrast with the French civil code. In addition, Pisanelli's civil code did not introduce divorce (Vincenzi Amato, 1988).

Liberal and lay forces strongly objected to the law on family relations included in Pisanelli's code. They first mobilized against the lack of provision for divorce. This was followed by protests against women's inferior legal status and the unequal treatment given to illegitimate children. Socialists and the working class supported their claims. At the same time, however, Catholics organized against any attempts to introduce divorce into the civil code. This was also later supported by more conservative forces and by parts of the Catholic women's movement. The Catholic women's movement also organized the acknowledgement of illegitimate children and better standards of living for women. Despite the various controversies, the law remained essentially unchanged (except for the abolition of husbands' authorization and the full access of women to every profession in 1919) until the new fascist penal and civil codes were approved in 1930 and 1942 (Saraceno, 1998). The lack of reform until the 1930s was probably due to both strong Catholic opposition towards the introduction of divorce and to fears that more equal distribution of power between men and women and the legal recognition of illegitimate children would compromise the unity of the Italian family. The same reasons might also explain the delay in family law reform during the Italian republican period.

The Take-Off Period, 1902–21

The origins of the Italian welfare state can be traced back to the last decade of the nineteenth century, as is the case with Germany, Austria and Belgium. In 1898 the first compulsory insurance scheme against industrial accidents was instituted, along with a state-subsidized voluntary scheme for old age and disability insurance (see Table 3). During the last two decades of the nineteenth century the 'social question' gained political attention and inspired both secular and religious groups to start making claims for state intervention. Progressive intellectuals and politicians began to denounce the extremely poor conditions of the Mezzogiorno (south of Italy). In the north, the first political mobilization began in response to the onset of industrialization and was followed by the foundation, in 1892, of the Italian Socialist Party. At the same time, in the *Rerum novarum enciclycas,* Pope Leo XIII encouraged a more advanced orientation in Church policy and placed the social question, especially the problem of the 'workers', at the centre of attention, moderating the Church's previously strong opposition to state intervention. The social doctrine of the Church was centred on the principle of collaboration and social harmony between classes and social groups and on the principle of Christian charity. In conclusion, then, we see that the Catholic and socialist movements, which formerly (and on different grounds) stood in opposition to state intervention started to ask the government for active intervention (Ferrera, 1984 and 1989).

Significant state intervention and a 'new social attention' became more concrete at the turn of the century. The new liberal leader, Giolitti, led the country towards democratization (universal male suffrage was introduced in 1912) and began the period of government intervention in the Italian economy and society.[3] The welfare measures of the Giolitti period led to a new 'parliamentary/democratic' approach to social intervention. They must be considered as answers to the new social demands coming from the 'bottom', and a tool of the new democratic competition.

Prior to the end of the nineteenth century, however, the Catholic Church retained almost complete monopoly over social policy. A large number of

Table 3
ITALY, DEVELOPMENT OF THE WELFARE STATE AND STATE POLICY TOWARDS THE FAMILY

	The take-off period (1902–21)
1898	Introduction of compulsory insurance against occupational injuries. Voluntary insurance against old age and invalidity
1902	*Female and child labour legislation
1910	*National Maternity Fund (Cassa Nazionale Maternità)
1919	Introduction of compulsory insurance against: *Old age (foreseen also for the widows and orphans) *Invalidity and unemployment (supplementary allowance in case of family dependent members)
	The fascist period (1922–43)
1923	*Discrimination against women's employment in public administration
1925	*ONMI (Opera Nazionale Maternità e Infanzia) was created
1927	*Tax for bachelors was introduced
1928	Introduction of tuberculosis insurance All employment contracts had to foresee protective health condition for the workers
1934	*Reform of women's labour law and maternity leave changes
1936	*Introduction of family allowances to dependent waged workers and employees
1937	*Family allowances were extended and scaled to the number of children in a family *Family loans were introduced and tax exemption was extended
1939	*Marriage and birth bonuses replaced maternity benefit *Health insurance for industrial workers (*casse mutue*) extended to family members
1940	*National Family Allowances Fund (Cassa Unica Assegni Familiari) was created *Family allowances were extended to spouses and to parents and parents-in-law with no or low income
1943	*General scheme of health insurance for private dependent workers and their family members

Note: Law related directly to the family are marked by an asterisk.

national and local charitable institutions provided basic poor relief. Confessional schools offered a widespread form of basic education (elementary school), especially in the centre and the south of the country (Ferrera, 1984). In 1890 a state law regulated the sector of poor relief, placing the religious institutions under state control.[4] However, while charitable institutions were put under the supervision of public authorities (mainly provincial authorities) for administrative and financial matters, no clear provisions were set to control the scope and the content of their action. Throughout the twentieth century the charitable institutions have carried out their actions in a state of great independence, acting as private bodies despite their semi-public status (Fargion, 1998). They have had an especially important role to play in family life, which has long been left in the nearly exclusive domain of the Church. The Catholic Church has traditionally intervened at the local level, in child care service provision, education policy, services for the elderly and basic poor relief. In fact, the church has regarded the state as a competitor and has thus resisted state attempts to assume responsibility for these areas.

In conclusion it can be underlined that in the first phase of its welfare state, Italy, like Germany and Austria, followed a 'social insurance model', granting entitlement according to individual workers' occupational status. In the second phase of welfare state development, that is, after the Second World War, Italy shifted towards an 'occupational mix' model.

The Initial Legislation on Child and Women's Labour

As illustrated in Chapter 2, during the last decades of the nineteenth century many European countries introduced their first state policies concerning the family or concerned with family issues. These were aimed mainly at diminishing the harmful effects of poverty and improving the living conditions of working mothers and their children. The health risks encountered by mothers and their babies immediately before and after childbirth were of special concern.

During the first decade of the twentieth century, only a decade later than in the more industrialized countries (e.g. Belgium and Germany), Italy also responded to concerns about working women and their children. The law on children's and women's labour protection was passed in 1902, while the first National Fund on Maternity Leave (Cassa Nazionale Maternità) was established in 1910. These were the first state policies concerning family responsibilities in Italy.

The 1902 law, strictly linked to the protection of child labour, was inspired by the idea that pregnancy and childbirth are two risky life events requiring protection. The law fixed minimum age limits and maximum working hours per day. It also regulated women's and children's labour during the night, in dangerous places and in heavy activities. Last but not least, the law introduced unpaid maternity leave during the first month after childbirth for female factory workers (Ballestrero, 1979). The 1902 law was then reformed in 1907 to bring it into line with the International Convention on Women's Labour signed in Bern in 1906. This law forbade night work for women.

As far as the support of the social and political actors is concerned, the 1902 law was the result of the mobilization of different political and social groups and in its final version contains a compromise of different interests. From the 1890s onwards, the Italian women's movement had placed requests and claims for maternity protection legislation within their wider campaign for suffrage, full legal equality and the right to hold office. Independent maternity funds were established in many Italian towns during this period with the support of local women's associations. These included the Lega per la Tutela degli Interessi Femminili (League for the Defence of Women's Interests) of Milan and Turin; the Associazione Generale di Mutuo Soccorso e di Istruzione fra le Operaie (General Association for Mutual Aid and Education among Working Women of Milan); and the Unione Femminile Nazionale (National Women's Union) (Buttafuoco, 1991). It should be mentioned, however, that the Italian women's political movement was crosscut by large divisions and debates as to the suitability and nature of state intervention in this sphere (Bock and Thane, 1991).

Italian employers strongly opposed legislation protecting women workers, especially when such measures increased the cost of women's labour. Some sections of the Italian feminist movement also opposed protective legislation, worrying that such laws would diminish women's employment opportunities. As a result, the final law was not passed until 1902.[5] In the end it represented a compromise between socialist interests and various government bills that protected, to some degree, the interests of industrialists (Buttafuoco, 1991; Ballestrero, 1996).

The second important step in women's labour legislation, and in state policies towards family responsibilities, was the 1910 law creating the National Maternity Fund for female factory workers. Both married and unmarried women who were obliged by the 1902 law to stop working during the first month after childbirth were entitled to an allowance which covered both childbirth and miscarriages that occurred after the third month of pregnancy.

The National Maternity Fund, especially in the years immediately following its introduction, met with much opposition and several obstacles. The main obstacle was governments' and employers' mutual refusal to pay for it. On the one hand, the government claimed that the state could not afford any expenditure on maternity funds, insisting that only workers and employers should contribute to them. On the other hand, the employers campaigned against the idea that the fund should be supported by themselves and their female workers without any state assistance. Finally, some parts of the women's movement also opposed the approval of the National Maternity Fund.

In its final form, the law required the state to subsidize the fund financed by working women's and employers' contributions. However, the provisions applied only to those women who were subject to the 1902 labour law (females working in factories), thereby excluding the vast majority of working women from any protection. The first compulsory maternity benefit was set up just as large numbers of women were being dismissed from the factories. Their exclusion from the labour force was the result of the Italian economic crisis of 1907, changes in industrial development, and, last but not least, the higher cost of female labour following the institution of protective legislation. Many women, as can also be seen in the British and French cases, were subsequently able to re-enter the industrial workforce at the beginning of the First World War (Pedersen, 1993). After the end of the war, however, employers and government reduced women's activity to ease social tensions surrounding male unemployment and to open up more positions to men. In the face of the heavy attack coming from the employers and without the solidarity of the trade unions, women did not have the power to resist mass expulsion from the factories.

The Issue of Family in the First Social Security Insurance Schemes

If the First World War signalled a break in social insurance development across all European countries, the turbulent post-war period marked a shift towards the approval of important social insurance programmes. In the period between 1918 and 1920 the liberal, socialist, and Catholic (*popolari*) political parties all presented several proposals and bills designed to introduce a health insurance scheme. A parliamentary commission on the reform of social security (Commissione parlamentare per la riforma previdenziale) was set up to this end in 1918. The commission's recommendations included unifying the existing old age, injury, disease, unemployment and sickness insurance programmes. It also advocated extending the scheme to all workers (including self-employed) and to their dependent family members. The recommendations of the commission were disregarded by the legislature. However, in 1919 two important compulsory insurance programmes were set up: the general scheme on compulsory insurance for old age and disability, and the unemployment insurance programme (Cherubini, 1977).

The insurance for old age and disability was extended to all dependent workers (blue- and white-collar, or *operai e impiegati*) between the ages of 15 and 65 and was placed under the supervision of the National Fund of Social Security (Cassa nazionale di previdenza). Regarding the issue of family responsibilities, this insurance scheme granted a monthly allowance for six months to workers' widows or orphaned children under the age of 15.

The first unemployment insurance scheme was also set up. Unemployment benefits were a pressing measure because of the grave unemployment problems following the war. The scheme was intended for male and female waged workers and for dependent agricultural workers (*braccianti*) involuntarily unemployed. The benefit also included a sort of supplementary family allowance (*aggiunte di famiglia*) granted to a spouse and children under age 12, and provided also for *de facto* family members (Cherubini, 1977).

To summarize, the take-off period preceding the fascist regime furnished four compulsory insurance schemes under the Italian welfare state: occupational injuries (1898), maternity leave (1910), old age and disability (1919), and unemployment for dependent workers (1919).

Before the advent of fascism, the issues of family responsibilities, caring work and family obligations had been tackled first through a maternity insurance scheme. Later on, family responsibilities were also recognized publicly through provisions for widows and orphans included in old-age insurance. Finally, the issue of family responsibilities would be acknowledged through the unemployment insurance programme, which provided a supplementary allowance for dependent family members.

The long history of the maternity insurance scheme illustrates the important role played by the women's movement, the trade unions and the progressive parties in claiming better working conditions and rights for working women. However, closer analysis reveals that many conflicting interests crosscut the various social actors. While trade unions defended working women's interests in the face of employer demands, they did not do so during periods of economic difficulty (the economic crisis of 1907, the post-First World War years) or high male unemployment. Likewise, male workers' interests and those of working women were not regarded as being the same.

The Fascist Period, 1922–43

The coming to power of Mussolini in 1922 marked a clear dividing line in the history of social policies relating to family responsibilities. The regime was deeply concerned with family issues and with the regulation of family matters. It was during this period that the foundations for the development of a family policy were laid. Indeed, fascist social policies would long remain (and still remain) the only systematic family policy that Italy has ever known.

The regime's main concerns about the family were related both to the fears of fertility decline, as in other European countries, and to the ideological support of a patriarchal model of family which approximated to that supported by the Catholic Church. Against the decline of the population the regime developed a set of welfare programmes and social policies which, given their comprehensive design, we can label 'family policy'. The ideological support to the patriarchal family was achieved mainly through giving support to fathers and fatherhood while weakening women's position in the official labour market.

The signing of the Concordat in 1929 may also be read as an attempt to instrumentally use the Church to reinforce male authority within the family and state authority (insofar as it represented the community's superior goals) over the family. The family was regarded, in fact, as an institution organically linked with the state, which one was supposed to become a component and instrument of. As Saraceno points out (1991), despite its great concern with the family as an 'ideological state's apparatus' (following Althusser's definition), the fascist regime did not substantially modify either dominant family values and behaviours or the role of the Catholic Church in regulating and allocating symbolic meanings and values. However, it is worth remembering that the fascist period witnessed several conflicts between the dictatorship and the Church, especially in the field of education. An illustration of this is the struggle in 1931 between the Catholic Association (Azione Cattolica) and fascist associations for the control of youth education (Di Febo and Saba, 1994). Although the Church and the fascist regime shared the vision that the family should be subject to the authority of a superior agency, they disagreed on whom that authority should be. Therefore, the Church and the regime both tried to exploit each other.

The Development of Family Policy

In terms of general welfare state development, the fascist period (1922–43) can be divided into two phases. The first one, from 1922 to 1927, which was characterized by the progressive repression of political debates and social conflicts in the political arena, experienced a limited expansion of state intervention through the welfare state, primarily in order to circumscribe the vigour of working classes. The second phase from 1927 (the date of the creation of the corporatist state) to the downfall of the regime in 1943, was marked by the systematic use of social insurance programmes as a means of social control. This prompted a great increase in welfare programmes, especially in the fields of sickness and family policy.

As far as social policies related to the family are concerned, the activity of the fascist government was particularly intense in the period 1932–37. The fascists' family policy package contained a wide set of social policy and welfare programmes, which ranged from legislation on women's labour, to child welfare and mothers' social services, to cash family benefits, such as family allowances, and fertility and marriage bonuses. Finally, fascist policies intervened in family matters through social security insurance changes, mainly through health insurance and survivors' pensions. Two main principles inspired fascist legislative activity regarding family responsibilities. First was that of the family wage, that is, the idea that wage levels should be related to family circumstances (ICSRI, 1943). Second was the inviolable role of the husband/father as family authority figure and breadwinner. Only single mothers or mothers married to disabled men were entitled to family allowances and bonuses at the birth of a child.

Maternity Insurance: Protection and Expulsion

During the fascist period legislation on women's labour underwent massive reformation and expansion. The origins of this legislation is found in both fascist ideology concerning the family and the political economy of the regime. As far as the first issue is concerned, it can be said that fascist discourse supported mainly a rural family model, characterized by men's marital and parental authority over wives and children. Women were seen as the producers of the 'race' and the pro-natalist policy was considered an essential component in contributing to the colonial expansion of the country. Therefore any experiences that might take women away from this task, from work outside the home to higher education, were viewed with suspicion (Saraceno, 1991). Women's work outside the home was not excluded in principle, but it was seen as acceptable only when necessary to complete the male breadwinner's wage (Ballestrero, 1989). At the same time, paid work at home or work in agriculture were seen as particularly fit for women, insofar as it did not remove them from the home and/or from under the marital authority of their husbands.

The political economy of the regime also influenced its policies concerning women's employment. One of the main problems faced by the regime was the growth of male unemployment. The legislation on women's employment during the regime aimed at discouraging women's labour, especially in the industrial sector and in the white-collar jobs of the newly expanding service sector. Thus, women's paid work was confined largely, albeit not exclusively, to the most marginal and unprotected sectors and positions. After 1923 measures aimed at cutting the number of women employed in public services and in schoolteaching were approved. In 1934[6] women's access to jobs and professions in public administration was made even more difficult. Gender-based discrimination was legalized.

Along with these discriminatory measures, the fascist government provided measures of assistance for pregnant mothers, newborn children and working mothers. In order to provide services to pregnant mothers and infants, the regime created the Opera Nazionale Maternità e Infanzia (ONMI) in 1925.[7] The ONMI, which lasted until the middle of the 1970s, consisted of several central and peripheral institutions catering mainly to poor and working-class working mothers and was inspired by two concerns. First was to help ensure the vitality of the 'race' by guaranteeing the health of pregnant and nursing mothers and their children. Second was to create an institution by which the regime could supervise and control how women approached their motherhood duties.[8]

Turning to protective laws for working women, the regime reformed the working mothers' protective law of 1902 in 1934. This reform[9] changed the maternity leave period from four to ten weeks while it established, for the first time, that a woman could not be dismissed from employment during pregnancy. The same law also changed the name from 'unemployment allowance' to 'maternity allowance' (Ballestrero, 1989). Finally, the law established that employers engaging more than 50 women were obliged to set up nursing rooms (*camere di allattamento*). In 1939[10] the maternity benefit for working mothers was abolished and replaced with marriage and birth prizes (ICSRI, 1943). This new benefit was not so much intended as an extension to the income support of mothers, as a support to fatherhood. Women were not banned from receiving it by law, but given the prevalence of male dependent workers over female workers they were *de facto* excluded. In general, this replacement and the very term 'birth prize' (*premio di natalità*) signalled a shift in attention from support to working mothers during pregnancy and childbirth to support for large families through a supplement to the breadwinner's wage.

Fascist legislation on women's employment can be viewed as illustrating that the dictatorship was more concerned with throwing women out of factory jobs than with protecting their health. As a matter of fact, the categories covered by maternity insurance, while it was still in force (1939), were not extended and most working women – home-workers (*lavoratrici a domicilio*), domestics, and women working in agriculture – were excluded. However, the fascist employment policies were not completely successful in sending women back to the home. The fall in the official female employment rate which only continued up until the 1930s, primarily affected the agricultural sector, and as discussed previously, was mostly due to legislation which favoured the employment of men. Despite the numbers recorded in official census data, most women worked as home-workers, servants, and/or in 'unprotected' jobs (Meldini, 1975).

In other words, despite fascist discourse in favour of creating a husband/father capable of maintaining his family, and despite discourses in favour of the model of the wife/mother, these arrangements were not compatible with a political economy that obliged women to work for pay. In fact, by 1936 the female share of the official labour force had returned to 1911 levels. While most women were occupied in 'unprotected' jobs, many did re-enter the official labour market.

Family Benefits

Fascist family benefits consisted both of direct cash benefits, such as family allowances and fertility bonuses, and indirect benefits, such as tax relief and tax exemptions. Among cash benefits, a fundamental role was played by family allowances, but fertility bonuses, marriage bonuses, and family loans were also important.

In Italy the origins of family allowances (*assegni familiari*) are found in a collective agreement signed by trade unions and employers' associations in 1934. The contingent objective of this benefit was to compensate working-class families for the reduction in wages which followed the work-time reduction from 48 to 40 hours (INPS, 1948; Hoffner, 1940; Franco and Sartor, 1990). However, as in other European countries, the introduction of family allowances in Italy was meant to reduce poverty, which was especially common among large families.

The Italian family allowance scheme had social insurance characteristics. Initially workers and employers paid contributions, and the family allowances were administered under the Social Insurance Scheme by the National Institute of Social Insurance (INPS), along with pensions, unemployment, tuberculosis, sickness and maternity insurance. The benefits were originally provided to male wage workers in the private industrial sector in respect of dependent children, and only later were they gradually extended to other sectors and categories. The law established a weekly benefit for each dependent child under 14 years of age.[11]

The pro-natalist purpose of the regime became more visible in 1937,[12] when it first increased the amount of each allowance according to the number of children. While family allowance originated as an anti-poverty measure, it soon acquired the status, beyond its pro-natalist objectives, of being one of the main instruments by which the state could implement the idea of the family wage. According to the law the following categories were eligible for the benefit: (1) heads of families (which the civil code established as being the husband); (2) widowed mothers, single mothers, or mothers married to men either unable to work or ineligible for unemployment insurance. In other words, women's eligibility was limited to cases where the father was absent or unable to perform his breadwinning duties.

In 1940[13] the benefit underwent important organizational and financial changes that would shape the characteristics of the scheme until 1988. First of all, a National Family Allowances Fund (CUAF) was created within INPS, with a separate budget and accountability. Next, employers were completely responsible for contributions to the fund. Finally, other dependent family members, such as spouses, parents, and in-laws with little or no income, became entitled to the benefit (ICSRI, 1943). The extension of family benefits to dependent family members other than children and spouses was a very specific feature of the Italian system of family allowances.

Besides family allowances, the regime introduced several other indirect family benefits and/or family taxes and direct benefits. In 1927 all bachelors aged between 25 and 65 were liable to a tax on bachelorhood. In 1937[14] in order to encourage the formation of new families, family loans as well as a supplementary allowance to unemployment benefit for workers with dependent children were introduced. In the same year, the regime passed a law enlarging tax exemptions and scaling them according to the number of children in a family.[15] This law also changed the level of a bachelor's tax and extended the obligation to members of the army. In 1939 the supplementary allowance for unemployment was modified and its amount was increased according to the number of children.[16] The Fascist Union for Large Families (Unione fascista tra le famiglie numerose) was founded in order to encourage large families.

Dependent Family Relationships in Social Security Schemes

Two other social security measures took into account the issue of family responsibilities: health insurance and survivors' pensions. In 1939 the existing and fragmented system of health insurance funds (*casse mutue*) was, by two national agreements, extended to family members of industrial insured workers (both blue- and white-collar). The provisions extended to the family included general medical and pharmaceutical cures, several assistance cures for childbirth, and coverage in case of hospitalization. As a consequence, the number of receipts from insured people and their family members in the industrial sector alone increased from about 750,000 in 1931 to about 7,000,000 in 1942 (Cherubini, 1977). Finally, in 1943 the National Institute for Sickness Assistance of Workers (INAM) was established and compulsory health insurance for all insured workers and their family dependent members was introduced.

In the same year in which health insurance was extended to workers' family members, Italy approved a law[17] which introduced survivors' pensions for widows, invalid widows, and orphans. However, the law stated that such measures would not come into force until the beginning of 1945. In the meantime survivors of deceased workers would receive a single allowance. In any case these laws set the stage for the future development of survivors' pensions.

Family Law Under Fascism

As I have mentioned, the fascist period did not produce a 'new' definition of family relations; it used the legal norms contained in Pisanalli's code of 1865 to define family obligations. The regime de-emphasized norms having liberal and individualistic inspirations and stressed the authoritarian principles of the regime and the righteousness of cultivating a 'common feeling' among the Italian people. In particular, the regime developed a concept of 'family interests', already promoted in 1910 by the Italian scholar, Antonio Cicu. The principle of the family interest broke with the 'contractual' and individualistic vision of the family linked to the liberal tradition, which was not fully incorporated into the 1865 civil code. The family was regarded as the organic basis of the corporatist society, charged with superior purposes, and centralized on the head of the family. The innovation was that the public sphere now ruled the private one. In 1935 Alfredo Rocco, father of the 1942 civil code, pointed out that the family had to be defined as a 'political and social institution', that is, family interests should be considered subordinate to those of the state.

Italy, however, changed its marriage law in 1929, when the regime and the Church signed a Concordat introducing the double matrimonial regime 'concordataire marriage' *(matrimonio concordatario)* that has remained unchanged until today. The Concordat was intended to solve the conflict between the state and the Church that had begun in 1870 following the conquest of the Pope's state by the new Italian monarchy.

The new penal code of 1930 produced a relevant modification to and articulation of different types of felonies that touched on family relations. Sexual violence and harassment were included in a section entitled 'Felonies against the public morality and the costumes' due to the 'public' nature of the acts (*titolo IX*). The abuse of correctional methods, bigamy, incest and adultery were defined as felonies against the family (*titolo XI*). Finally, abortion and contraception were defined as felonies against 'the race' (*titolo X*), and remained as such until 1978, when a new maternity law on the voluntary interruption of maternity was enacted.

Rocco's code of 1942 only slightly changed the previous norms of family relations. Family and kin member obligations were maintained, as elaborated in the 1865 code, but they were made into a separate section of their own (*libro I, titolo XIII, 'Degli alimenti'*). According to legal obligations, family and kin members were obliged to provide financial support for each other in proportion to their income and their degree of kinship. Under Italian fascism 'the family was subject to pyramidal and complex rules' based on the authority of the husband/father but also responsible towards and under the guardianship of the state and the regime (Pocar, Ronfani, 1998, p. 28). Innovations related to public control over children in public and private assistance can also be read in this light. The law established that a child could be placed in the custody of a tutelary judge (*giudice tutelare*) or other person who had taken care of him or her for at least three years, through the institution of affiliation (*affiliazione*).

Most of the principles contained in the penal code of 1930, and also especially the civil code of 1942, continued to regulate family relations in the new democratic Italian republic until in 1975 family reform law modified the section concerning the family.

NOTES

1. In Wallerstein's thesis the whole 'world economy' can be interpreted in different historical phases, in terms of countries belonging to a small 'core' zone, a larger 'semiperipheral' zone, or an even larger 'peripheral' zone. What has to be emphasized in this vision is the fact that because of the spatial hierarchization inherent in global capitalism, there are different political processes at the state level. From this perspective a shared geographical position (i.e. Mediterranean countries) has been an important factor in producing a pattern of political–economic convergence (Wallerstein, 1979).
2. Here the industrial revolution and the origins of the modernization of Italian and Spanish societies cannot be analysed in depth. A wide literature exists for the Italian case. See in particular De Rosa (1980), Cafagna (1961) and Merli (1972). For the Spanish industrialization process see Soto Carmona (1989) and Giner (1990).
3. An increase in total public expenditure accompanied the introduction of new laws and welfare measures, with spending passing from 13.6 per cent of GDP in 1900 to 14.7 per cent in 1913, reaching the levels of other European countries, such as Great Britain (12.7 per cent), and Germany (14.2 per cent). On this issue see Flora, and Heidenheimer (1981a).
4. For an analysis of the development of public assistance policies in Italy from 1861 to the fascist regime, see Fargion (1983, 1998) and David (1984).
5. Law no. 242, 19 June 1902, known as Carcano's law.
6. Law no. 221, 8 January 1934.
7. Law no. 2277, 10 December 1925.
8. For a detailed report on ONMI's activity, see Saraceno (1991).
9. Law no. 1347, 5 July 1934.
10. Regio Decreto Legge (RDL) no. 636, 14 April 1939.
11. RDL no. 1632, 21 August 1936, converted into Law no. 404, 18 January 1937.
12. RDL no. 1048, 17 June 1937.
13. Law no. 1278, 6 August 1940.
14. RDL no. 1542, 21 August 1937.
15. RDL no. 463, 4 February 1937.
16. RDL no. 636, 14 April 1939.
17. Law no. 1272, 6 July 1939.

Chapter 4

THE SPANISH CASE

This chapter looks at the Spanish case from the establishment of the first women's labour law to the end of the Franco regime. Given the long time span of the analysis, I have divided the case into three main periods. The first, the 'take-off period' (1900–30), extends from the implementation of the first women's labour law to the advent of the Second Republic. The next period, the 'Second Republic' (1931–36), is dedicated to providing an overview of the break in law and in social policy that this brief epoch represented in relation to both past and future Spanish dictatorships. Finally, the analysis continues with the illustration of the third period, which lasted up to the democracy: the Franco regime. The Franco period (1939–75) can be further broken down into two subdivisions, marked by a change in economic, political and social conditions. The first period, 'the period of autarchy' (1939–59), lasted until the end of the 1950s. The second period, the 'capitalist period' (1959–75) ended with the demise of Franco. This breakdown is also justified by a change in political orientation of the dictatorship concerning social policy.

The Socio-Economic Context

A general picture of the socio-economic context in which the first social policies related to the family were embedded is provided here with the use of two main indicators: labour force by sectors, and women's economic activity. I later try to link the evolution of the main social policies as well as the foremost women's labour legislation and maternity insurance to these general developments. I wish to look at how the occupational structure in Spain (and Italy), characterized by low rates of female activity in the official labour market, has shaped public family provisions and formal caring services. The analysis tries in part to answer the following questions: what have been the main consequences of the low proportion of female waged industrial workers? and how could the male breadwinner model be achieved in a deeply agrarian society where the devotion of women to unpaid family work alone was economically impossible for many families?

Occupational Structures

Over the last century the changes in the structure of the labour force in Spain have been marked by three main trends, some of which are common to other western countries, others of which are rather more specific and/or common to the other southern countries. The first trend, common to other industrialized countries but later in timing, has been a steady and large reduction of the active population in the agricultural sector. Specific to the Spanish case is the fact that it underwent a new process of ruralization after the civil war. The total active population in industry was 27.5 per cent in 1930, a figure achieved again only at the end of the 1950s. The reduction in the agricultural labour force became particularly deep only after the 1950s. The second trend, which differentiates Spain from other industrialized countries, but also from Italy, is that the increases in the industrial sector labour force outpaced those in the service sector only up until the 1960s. Since then Spain has experienced a concentration in demand from the third, service sector. The 'tertiarization' process is connected to the development of new consumption patterns, the growth of demand in the public sector, and specialization in the tourist industry (Hernández, 1990).

Finally, as in the Italian case, the third main labour force trend of the last decades of the twentieth century has been a decline in the activity rate and an increase in unemployment rates, in particular because of women's recent entry into the labour market. This phenomenon is coupled with the growing importance of the 'grey' or 'informal sector' of the economy.

Figure 3
LABOUR FORCE BY SECTOR, SPAIN, 1877–1981

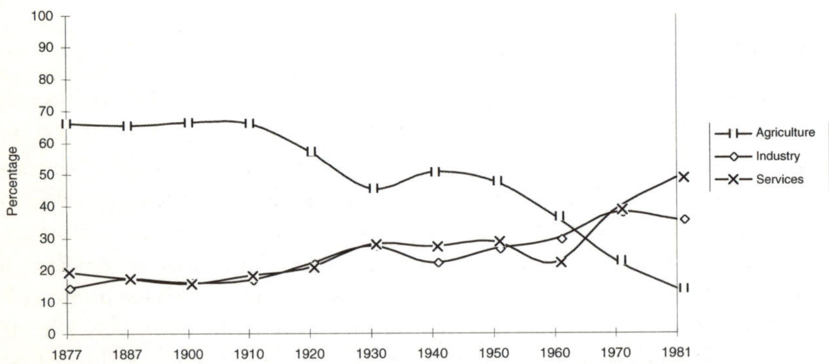

Source: Based on figures provided by Carreras (1989), p. 79. Due to insufficient data women's activity in agriculture is not included.

Female Labour Force Participation

Female labour force participation in Spain during the twentieth century experienced considerable growth, the form of which approximates to a 'u-shaped' curve. Data available for the official labour market show that the first period between 1900 and 1940 was characterized, as in the Italian case, by a decline in the female share of the labour force, which dropped from 18 per cent in 1900 to 12 per cent in 1940. The main reason for this decline is to be found in the decreasing importance of the agricultural sector, which had been a main employer of women (López, 1988). The agricultural sector lost about 428,000 working women from 1900 to 1930 (Espina, 1990). Some of these women assumed low-skilled jobs in the industrial sector, mainly in textiles. In fact, according to Nash (1983), the percentage of wage-earning women in the industrial sector rose from 16.2 per cent of all industrial workers in 1900, to 31.8 per cent in 1930. Homework was, however, the prevalent form of employment for women. At the beginning of the century 50.8 per cent of women working in the industrial sector were homeworkers. By 1930 this percentage had dropped to 25.6 per cent of all wage-earning women in the industrial sector.

Just as in the Italian case, it is also important to consider that the very low figure for women's official activity in the Spanish labour market is partly due to the problematic way in which census data were collected until

Figure 4
FEMALE SHARE OF THE LABOUR FORCE, SPAIN, 1900–80

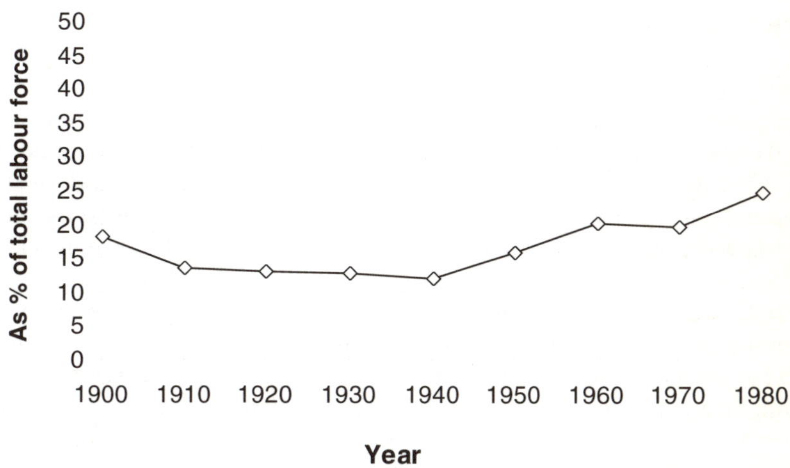

Source: Based on figures supplied by Carreras (198), p. 77, table 2.14.

the 1964 Survey on Active Population (EPA).[1] Spanish women's activity rate in the official labour market, like the rates of women in other European countries, has grown constantly since 1950.

The level of female employment and its variation across countries does not simply reflect the pattern of dependence produced by labour market structures, but is also influenced by dominant ideas about women's more appropriate role in society and in the family. Nash (1983) provides information concerning prominent attitudes towards women's employment, drawn from texts and documents covering a period from 1875 to 1936. She argues that the main conservative and Catholic-oriented thinkers since the end of the nineteenth century have regarded women's employment, especially outside of the home, as not fit for women and as a danger to the family. But given that most women were working for pay, this hostile attitude towards women's employment weakened and became mitigated by the belief that women's paid employment was acceptable in cases of economic need. During the same period, however, some left-wing thinkers expressed quite different attitudes towards women's employment: women's access to work must not be justified by the miserable conditions of family life, but rather conceived of as a woman's right. Not all left-wing thinkers shared this opinion, however, especially during periods of economic crisis. The future promoter of the Spanish Communist Party argued that within the labour movement women's employment was a terrible source of conflict with male employment and with male wages.[2]

The Legal Context

The Roots of Spanish Family Law

The roots of contemporary Spanish family law can be traced back to the civil code of 1889, on which the Napoleonic code of 1804 had a major influence (Tomas y Valiente, 1992). The 1889 civil code was a portrait of the conservative Spanish society of the end of the century, and was based on a bourgeois order of society; upon a starkly unequal distribution of rights and duties between genders and across generations. Husbands were the legal administrators of the common wealth produced by the income of both spouses, and they administrated wives' property. Women were totally subordinated to their husbands' authority in the disposal of their property or for acting in a judgment. Adultery was considered by the penal law as a crime, but men and women received different treatment for extra-marital relationships. Parental authority was vested in the father and could only be assumed by the mother in the event of his absence or death. Although legal adulthood was set at 21 years of age, daughters under the age of 25 needed parental permission to leave their home, unless they were getting married. Finally, the civil code established the defence of legitimacy as one of its main principles; therefore, paternity investigation was forbidden.

Two features marked the marriage bonds of Spanish family law: unity and indissolubility. Spanish society at the end of the nineteenth century would not accept the divorce law of the Napoleonic code; in this respect it was similar to the Italian civil code of 1865, which adopted many principles inspired by Napoleon but not divorce law. Spain also maintained both the canonical and civil marriage system in order to keep a good relationship with the Church and to satisfy the conservative bourgeoise. Canonical marriage was compulsory for every couple in which at least one of the marrying partners was Catholic. In cases where a baptized Catholic wanted to marry under civil law, it was necessary to produce a declaration of apostasy. The Spanish civil code of 1889 refers to the nuclear family where the essential relations are between couples and their offspring, bolstering a process of emancipation from a wider circle of kinship. The couple and the children constituted the core of the bourgeois family, though legal obligations were still defined broadly enough to include relatives outside of the nuclear family.

Comparing the main influences of the legal contexts of Italy and Spain, it is worth underlining that with the exception of divorce, family law in both countries has been influenced by the Code Napoléon. In addition, both countries upheld concordats between the state and the Church that influenced the legal definition of marriage for many years.

The Take-off Period, 1900–30

The origins of the Spanish welfare state are to be found in the governmental reforms (Comisión de Reformas Sociales, 1883) put forward at the turn of the nineteenth century. The introduction of the initial welfare schemes in Spain was closely linked to the influence exercised by the thoughts and ideas of two different groups of intellectuals: *'neo-liberal' krausistas* and *social Catholics*. Both of these groups, for instance, are said to have played significant roles in the creation of the National Institute of Insurance (INP; the Instituto Nacional de Previsión Social) in 1908 (Moreno and Sarasa, 1993).

The 'neo-liberal'[3] *krausistas* intellectuals, inspired by the German philosopher Karl Krause, fostered a programme of promoting a mutually beneficial coexistence of classes which would constitute an alternative to traditional laissez-faire conservatism, Catholic corporatism and revolutionary collectivism. They believed that the main cause of poverty in Spain was its cultural and economic 'backwardness'; therefore, the country should improve its economic and social structures by means of raising education levels.

The social Catholic groups, which represented a minority sector of Spanish Catholicism, argued for greater social justice based on the idea that not all social inequalities have a natural origin. Some of them sought to promote trade unions to protect workers against employers. Their general focus was, however, on the moral regeneration of the poor through piety rather than by means of economic and social reforms (Moreno and Sarasa, 1992). Guillén (1996) shares to a great extent Moreno and Sarasa's approach. This author argues that the ideology and perceptions of the political and intellectual elite are important explanatory factors in assessing the origins and development of the Spanish welfare state.

The Initial Legislation on Women's Labour

In Spain the first legislation on women's labour stretches back to a law of 1900[4] that established a rest period for mothers for three weeks after childbirth. Subsequently, in 1907[5] this first law was reformed and the compulsory leave period was extended to four weeks. However, these first measures, despite their explicit goal to protect working mothers, actually intensified poverty amongst women who were not awarded any pay or allowances during the compulsory leave period.

In the field of social insurance measures, the INP promoted a private and voluntary system for the payment of old age pensions and accident insurance during the first decade of the twentieth century. With the unfolding of the revolutionary movement, which resulted in the general strike of 1917, the Spanish government was confronted with the limits of the existing social insurance schemes. It subsequently proposed new universal insurance schemes covering the risks of old age, sickness, unemployment and maternity. Employers and landowners were very hostile towards the proposals, with the latter group supported by a large sector of the Catholic Church. At the same time, doctors' associations and insurance companies strongly opposed the introduction of compulsory health insurance. The working-class organizations were not well organized and did not articulate a united front.

These insurance proposals were then stopped by the advent of the dictatorship headed by General Primo de Rivera (1923–30), who was supported by traditional oligarchic groups. However, the first compulsory social insurance scheme for pensions (*retiro obrero obligatorio de vejez*) had already been introduced and legislated by 1921.[6] In addition, the Rivera dictatorship introduced a maternity benefit in 1923[7] (Cuesta Bustillo, 1988; Nuñez Perez, 1994). The maternity law of 1923 established a maternity allowance and medical assistance during pregnancy for those working mothers who were paying contributions for old-age insurance. At the same time the INP was elaborating a new project for the implementation of a maternity scheme, which was approved some years later.[8]

The Second Republic, 1931–36

The INP's plan for a unified social insurance scheme was launched again during the Second Republic (1931–36).[9] Despite republican policy-making efforts to establish a unified system of compulsory social insurance, the only new scheme introduced was that of insurance against occupational injuries (1932). Not only had plans for a universal insurance scheme failed, but insurance against unemployment and illness also remained out of the compulsory system. The military uprising of Franco and the subsequent civil war destroyed any further efforts.

From a legislative point of view, women experienced great improvements in their position in society during the Second Spanish Republic (1931–36). Women achieved suffrage, divorce was introduced, and most discriminatory legislation in the areas of family jurisdiction, politics and work were repealed. In practice, however, the achievement of equal social rights remained out of reach.

The Maternity Insurance Scheme

The maternity insurance scheme of 1923 originated within an international context marked by the end of the First World War, by women's claims for political and social rights, by the Conference of Washington, and by a general trend towards the enlargement of insurance schemes among the majority of European countries. The law that enacted a wider compulsory maternity insurance scheme occurred only some years later, in 1931, during the republican government.[10] One should note that the Rivera regime had elaborated a project on maternity insurance prior to the 1931 republican government and the above-mentioned law. The date of implementation for this project, however, had not been reached (Cuesta Bustillo, 1988; Nuñez Perez, 1994).

The maternity insurance scheme of 1931 was aimed at assisting insured workers through medical assistance and cash benefits. It would allow working mothers to leave their job during the period around childbirth and favour their access to the Obras de Protección a la Infancia y la Maternidad (OPIM). This latter scheme of the Spanish government provided services such as medical assistance centres for pregnancy, childbirth and infants. The OPIM also controlled maternity leave and provided information on infant and child care to mothers, much like the ONMI, which was created by the Italian fascist regime.

The beneficiaries of maternity insurance were dependent women workers between 16 and 50 years who were already insured for old age. Domestic workers remained excluded. The maternity leave period was fixed at six optional weeks before birth and six compulsory weeks after delivery. There were five different types of benefits: (1) medical and pharmaceutical assistance; (2) cash allowances for the compulsory and the optional maternity leave period related to the period of the insured contributory records; (3) free utilization of services and medical assistance provided by OPIM; (4) allowances for breast-feeding with a fixed maximum period of ten weeks; (5) supplementary allowances in cases of special problems. The financing of the insurance scheme was assured by employers' and workers' contributions and by subsidies from the state. It was administrated under the supervision and the organization of the INP. However, in the following months after implementation the ministry passed orders to reduce eligibility. It fixed a minimum contributory period towards old-age insurance prior to eligibility for maternity benefits. Therefore, the majority of working mothers, who typically held precarious or seasonal jobs were excluded (Nuñez Perez, 1994).

Despite the new restrictions, coverage greatly increased from 500,378 in 1932 to 741,771 in 1935. Most of the beneficiaries were industrial dependent workers (63.1 per cent), with the wide variations among regions reflecting the differences in industrial development within the country. Even though the number of beneficiaries increased, only about half of insured women had completed the minimum contributory period necessary to receive cash benefits during the leave. This figure points to the fact that precarious and seasonal jobs were particularly widespread among women (Nuñez Perez, 1994).

Despite wide concern about poverty, infant mortality and the labour conditions of working mothers, Spanish political and social leaders also raised opposition to the introduction of maternity insurance. Opposition towards the law stemmed chiefly from the employers and partly from workers. The main resistance among the working class came from working women who paid the contribution for the insurance but did not hope to benefit from it, such as single women, widows and childless women (the opposition coming from Catalonia was especially strong). The strikes and protests were often supported by the CNT (Confederación Nacional del Trabajo, the anarchist trade union). They lasted only a few months, when the legislation was first introduced. After about a year, most working women realized the greater advantages that they would gain from this insurance.

Campaigns and propaganda were used to counter the initial resistance to the law, especially between 1931 and 1932. Several associations linked to the Unión General de Trabajadores headed these campaigns. The INP also organized campaigns, informative papers and conferences in support of the insurance. The explicit goal of such campaigns was to help female workers understand how the new insurance scheme functioned and how they might benefit from it. However, the arguments utilized during the campaigns clearly suggest that women's extra-domestic work was not always looked upon with favour. Most of the actors who voiced support for the new maternity insurance scheme did so less out of a conviction that women should work than out of a desire to minimize the social evils of women's paid employment. During the periods of economic crisis, the socialists also held that the ideal family model was the one in which women were homemakers and men were breadwinners. They regarded women's paid work within the context of home-making, such as domestic work, as acceptable and even desirable (Nuñez Perez, 1994).

According to Cuesta Bustillo (1994), the process by which maternity insurance was implemented in Spain constitutes one example in women's history where women themselves were absent. The fact that the women subjected to maternity insurance law did not play any role in either the elaboration or the implementation of the law might be explained by the absence of organizations advocating for working mothers. Instead, the government and legislature charged the INP with designing and carrying out the project. Since 1932, the INP drafted provisions for maternity into the unified social insurance scheme. Medical assistance and cash benefits were included in health insurance, and the provisions were extended to employees' wives. The advent of the civil war stopped any further extensions.

The Franco Regime, 1936–75

The Franco regime stemmed from a heterogeneous alliance of conservative and fascist groups. It was founded as a reaction to the republican phase, and it was supported by a *connubio*, or partnership, between the army, the big landowners, the Church, and traditional conservatives (Salvadori, 1982). Put in a nutshell, it can be said that 'Francoism was above all representative of a reactionary despotism which aimed to accelerate a rapid process of capitalist accumulation' (Moreno and Sarasa, 1992: 6).

Although a few, very limited measures relating to the protection of the family already existed before the civil war (1936–39), the development of family policy in Spain was strictly linked to the advent of Franco's dictatorship. The ideology of the new regime invoked the greatness of Spain as based on imperial expansion – namely on fascist-corporatist and traditional Catholic values surrounding the family. That ideology, along with an overriding concern about depopulation following the decimation of the civil war, led to the development of pro-natalist policies. The ideology and political economy of the Franco regime also converged in the development of family policy and in measures aimed at supporting family responsibilities and dependent family relationships. Restoring the traditional Catholic family model and controlling wages were the first and second impetuses behind family policy.[11]

There are two movements that developed during Francoism which are particularly relevant to the analysis of the development of family policy and social policy related to family responsibilities, especially during the early phase of development: National Syndicalism, and National Catholicism (Nash, 1991). The former was influenced by the fascist-corporatist ideologies of *Falange* (Moreno and Sarasa, 1993). The regime claimed to be an alternative to both liberal capitalism and Marxism and deployed direct interventions in the economy through what has been called 'vertical syndicates'. Every branch of the economy was organized in vertical syndicates, the membership of which was compulsory for employers, administrative staff and workers. The worker was regarded as a unit of economic production and was entitled to receive social protection against unexpected risks. The wage system during this period was totally under the control of the regime.

The other movement that was relevant to the development of family policy was national Catholicism. Traditionalists supported this, and it played a central role in reinforcing the traditional role of the family and women's complete subordination within it. 'Our State must be Catholic in the social' stated Franco in 1937 (Moreno and Sarasa, 1993, p. 7). This Catholic paternalism as elaborated by the most reactionary sector of Spanish Catholicism, is reflected in the fact that social policy during the regime was called *obra social* (social task). National Catholicism represented the politicization of the Church's traditional pro-natalism, familialism and acquiescence to female subordination. As a matter of fact, one of the first requests the Church made to the new regime was that its control over the family and marriage laws be reinstated.

The influence of National Catholicism, National Syndicalism and the pro-natalist aims of the regime led to the restoration of strong familialism. The family was regarded as a natural society, a primary social unit, a basic cell in the body politic of the state and the community. The status of the family as the prime unit of social organization was complemented by declarations about the indissolubility of marriage. The regime combined pro-natalist policies with the family's consolidation and material support. As a result, pro-natalist rhetoric and familial ideology was interwoven with family policy. Women were particularly affected by the new regime, which advocated stable Catholic families with a strong patriarchal authority at the head. To this end family subsidies were introduced within the social security system to encourage women to remain at home. Along these lines, male dependent workers were awarded benefits for dependent family members according to the principle of *carga familiar* (family charge). Furthermore, the Franco state encouraged the 'return' of women to the home. Given that social security provided only for dependent industrial workers, women outside of the family sphere were without any social protection. Children who did not have an insured father also remained without social protection. During the Franco regime social rights did not belong to children themselves, but to their married fathers. Finally, largely in response to past republican propaganda, the regime restricted women's legal rights, subordinated their roles completely within the family, and severely defined their moral behaviour according to the model given by the Church.

The Period of Autarchy, 1939–59

From a political and economic point of view, the early years of Franco's regime were characterized by a strong corporatist and autarchic orientation that excluded any foreign interference. During this period social policy was largely neglected and social protection was contingent on both charity and beneficence. Specifically, the social insurance system offered a restricted array of benefits and services to a small proportion of the population, which was composed mostly of low-income industrial workers and their dependent family members (Guillén, 1992). In 1939 the state introduced old-age subsidies and disability insurance for low-income industrial workers. In 1947 this plan was transformed into the Old Age and Invalidity Insurance Scheme (Seguro de Vejez y Invalidez, or SOVI). It established a fixed pension for low-income industrial workers, and had to be co-financed by the employers and the state.

The development and the growth of old-age insurance was slowed and partially replaced by the creation of mutual aid associations (*mutualidades laborales*). The *mutualidades laborales* was a peculiar institution in the Spanish social insurance system. It worked in parallel to and concurrently with the public social insurance system, providing benefits to workers in every sector of activity at national or provincial levels (Cruz Roche, 1984).

In 1955 a measure related to family responsibilities that recognized women's dependence on the breadwinner was established through the introduction of a sort of initial form of survivors' pension. This policy was considered a new extension of the SOVI, and extended benefits to poor, elderly widows of insured workers. While the Spanish social insurance system was very poor throughout this period, offering few and minimal benefits to a small percentage of the population, measures directed at families largely expanded. The state developed strong support for the male (industrial worker) breadwinner's family responsibilities, making it one of the main pillars of social policy. According to Meil (1994), family benefits (that is, direct family transfer payments) constituted more than 50 per cent of total public expenditure on social security in Spain from the 1940s up to the end of the 1960s. The high level of expenditure on family benefits can be explained by the lack of other types of social protection and by the use of family policy to both restore the traditional family model and to balance the effects of wage policies.

Family benefits The first general scheme of family allowances (*subsidio familiar*) was introduced by the dictatorship while Franco was still waging civil war in the occupied regions, in 1938. This scheme, like that introduced in Italy, was mainly intended to alleviate the widespread effects of poverty, which were especially common among large families. The Spanish *subsidio familiar* had the character of a social security benefit and was provided to most male wage workers but denied to domestic servants and home-workers. The benefit consisted of a monthly cash benefit paid directly to the *jefe de familia* (head of household), with at least two children. The pro-natalist purpose of the benefit in the Spanish case was visible since its introduction by the fact that it was provided only to families with at least two children and by the fact that the amount of the allowance increased with the birth of subsequent children.

The allowances were co-financed by contributions from workers and employers. From 1941 onwards the state also contributed to the financing of the *subsidio familiar*. According to the rules of the scheme, the benefit was explicitly for families with children. Despite this, the scheme quickly expanded to include other beneficiaries. Insured dependent workers were entitled to the benefits not only for their economically dependent children, but also for dependent grandchildren and siblings under the age of 14 (18 if they were in school) whose fathers were dead or unable to work. Beneficiaries had to be living with and supported by the insured worker. These same categories of extended family members were also eligible for benefits if they were over the maximum age but were permanently unable to work. This extension of benefits beyond the 'nuclear' family is a peculiar feature of state policy towards family responsibilities in Italy and Spain. Eligibility for the *subsidio familiar* was extended also to the widows and orphans of the insured worker.

Since 1942 the spouse of an insured worker who was entitled to *subsidio familiar* had the right to use the existing maternity insurance services, given that maternity insurance had been repealed and replaced with health insurance. In 1941 other family policy measures had been introduced. Loans, marriage subsidies and yearly prizes for large families encouraged marriage and reproduction. Large families were entitled to many types of fringe benefits, ranging from transport and school grants to tax exemptions, credit and housing and sanitary assistance (Meil, 1994; Nash, 1991; Rull Sabater, 1971).

In 1943 agricultural workers also became entitled to *subsidio familiar* by a special social security scheme. The excluded groups, as in the Italian case, were the self-employed and domestic workers.

The wide range of family policies established during this early period of the regime were enacted during a phase of economic difficulties, scarce availability of public resources, and a high level of inflation. In this context they portrayed their limits as protective measures for families. First of all, the amount of the *subsidio familiar* was minimal and not indexed to prices. At the same time, the funds constituted by the contributions were often used in order to create new insurance schemes (Meil, 1994). The amount of the allowances increased until 1945, but from then onwards remained frozen (Nash, 1991). During the mid-1940s the low level of public expenditure for social security and a high level of inflation helped to expose the limits of the protective functions played by the existing system of family policy. Wage control, coupled with the *subsidio familiar's* low level of protection, sent a large number of families into poverty.

In order to improve the social conditions of families, the government introduced a supplementary wage system, *plus de cargas familiares* (family bonuses) in 1945. Not an insurance scheme, the *plus de cargas familiares* instead complemented male industrial worker salaries, and can be regarded as a sort of 'family wage'. This system, from a comparative perspective, had not, at that time, an equivalent and can be defined as unique to the Spanish family policy system (Meil, 1994). It was a very particular measure for two reasons. First, its method of functioning. Firms created funds on the basis of the wage volume of the enterprise. These funds were then distributed among the married wage-earners according to a points system. Married couples were awarded 5 points on a sliding scale which went from 6 points for one child to 30 points for ten children and 5 more points for any other child (Rull Sabater, 1971). During the 1940s it constituted around 5 to 20 per cent of the total wages of any specific firm, while during the 1960s the percentage was around 25 (Meil, 1994). The percentage of contribution varied according to the sector of activity of the enterprise, and even then the value changed from enterprise to enterprise and over time. The second unique feature of *plus de carga familiares* was the extended notion of dependent family members on which it based its criteria of inclusion. However, it has to be noted that although *plus de carga familiares* was soon extended in order to widen the number of eligible dependent family members, it was designed mainly to support families with children. The following categories were entitled to *plus de carga familiares*: male married workers; male widowers with children; workers economically responsible for elderly (over 60) or disabled ascendants, and workers responsible for disabled or orphaned siblings (under 14) (Rull Sabater, 1971).

The *plus de carga familiares* met two of the regime's requirements. First, it met the desire to create a real 'family wage' and to support families with children and the male breadwinner family model. Second, it fitted well into the wage policy of the system. The *plus de carga familiares* supported the male breadwinner family model because it was a measure paid to the father and aimed in the first place to encourage women to stay at home. As a matter of fact, the preamble of the decree of the *plus de carga familiares* states that this wage supplement has to be regarded as a first step in the direction of a family wage and thus it was intended to bring married and employed women back to the household. Working mothers were penalized as married couples received no monetary compensation if only the wife worked. The *plus de carga familiares* was paid to the wife only in very exceptional circumstances.

Domestic servants, family workers (sectors in which women were mainly employed) and, more importantly, agricultural workers were excluded from the *plus de carga familiares*. Over 50 per cent of the active population was still employed in the agricultural sector in 1940, and so the majority of workers were excluded from this system. This fact constituted a grave shortcoming in the Spanish general system of family protection, especially since the real value of *subsidio familiar* was declining. However, from the 1950s until the 1960s the *plus de cargas familiares* became the main instrument of family policy for non-agricultural workers in Spain – so much so that other social categories also claimed the supplementary wage. Finally, in 1954 the government created an ad hoc family allowance for civil servants.

Health insurance Health insurance (*Seguro Obligatorio de Enfermedad*, or SOE) was introduced in Spain in 1942.[12] It originated in the voluntary schemes previously run by medical societies and insurance companies, as well as in the support offered by the employers in this field (*sociedades médicas y compañías de seguros*).

The institution of health insurance marked the end of the maternity insurance scheme, which became subsumed under the health insurance scheme. The new law established that insured working women were entitled to an allowance (a cash benefit) during the compulsory maternity leave, for a maximum of six weeks. During this maternity leave women had the right to 60 per cent of their normal wages. There were special grants in case of multiple births. The suppression of the maternity insurance scheme occurred along with the introduction of a set of women's labour and protective laws that aimed at limiting women's participation in the labour market (Alonso Olea, 1994).

Spain developed its first health insurance scheme to cover only a small part of the population with the provision of limited cash and health benefits. As with other social insurance measures, the right to health insurance was directed to low-income industrial workers and their dependants. Only the insured worker had direct access to the cash benefits, while the dependants of the insured were entitled only to the health services. Eligible family members included spouses, dependent direct descendants and siblings under the age of 18 or unable to work and dependent ascendants. Wives were eligible independently of their economic position in the labour market and were able to maintain their beneficiary status even in the case of legal or *de facto* separation. In contrast, the law stipulated that all other beneficiaries must be economically dependent on the insured worker. Dependent family relationships were symbolized by the creation of the *cartilla* (a family document). According to the *cartilla*, the insured worker was entitled to health insurance for himself and for his family. The use of the *cartilla* survived in the Spanish health services beyond the end of Franco's regime. This wide extension of the number of beneficiaries of health insurance is another example of the inclusion of a broad family definition in social security measures (Alonso Olea, 1959; Rull Sabater, 1971). In 1950 only 29.4 per cent of the population was covered by the insurance; the figure rose to 43.3 per cent in 1960 (Freire, 1993). Autonomous workers in the agricultural sector were one of the larger groups excluded from health insurance.

The Capitalist Period, 1959–75

From an economic and political point of view the period between 1959 and 1975 corresponded to that of economic 'developmentalism' (Moreno and Sarasa, 1993). In 1959 the Plan of Stabilization was approved, marking a clear break from the progressive liberalization of the Spanish economy. Some further steps were taken in order to establish a system of labour regulation. Since 1958 a Collective Agreements Act allowed employers and employees to negotiate wages at the firm's level.

The 1960s brought major changes in the political orientation of social policy. In an attempt to integrate the country into the international economy, the Franco regime adopted a new deal that tried to combine an authoritarian political system with Keynesian demand policies. Although this new deal was initiated as a means to spur capitalist modernization, it resulted in the opening up of a new wave of social reforms led from above, completely changing the path of social welfare in Spain (Rodríguez Cabrero, 1989). Two main reforms marked the transformation of social welfare during this period. First was the Basic Law of Social Security (Ley de Bases de la Seguridad Social, or LB) of 1963. Second was the 1972 reform, the Ley de Financiación y Perfeccionamiento, or LFP, which can be considered as the first attempt to reform the fiscal system. The LB, which was not implemented until the second half of the 1960s, can be considered the first move towards the foundation of welfare 'universalization'. It unified the previously existing schemes into a single institution, the National Social Security System, and eliminated eligibility requirements that made the system accessible only to low-income workers. However, because financing of the system continued to be based on contributions from the affiliates, it continued to be highly fragmented, with participation based mainly on professional lines (Guillén, 1992). The 1972 reform, the LFP, intended to expand social protection especially in the areas of pensions and unemployment benefits. As a result of these reforms, public social expenditure increased from 36.2 per cent of total public expenditure in 1960 to 56.6 per cent in 1970 and to 62.1 per cent in 1975 (Rodríguez Cabrero, 1989, table 3). In the 1960s social expenditure in Spain was very low in relation to GDP (7.3 per cent in 1960, 10.6 per cent in 1970, and 12.9 per cent in 1975), but since then there has been a change in the dynamics of growth and the size in relation to public expenditure (Rodríguez Cabrero, 1989, table 2).

The unification and reorientation of family policy In the context of the general social policy reforms begun in 1963, family allowances were unified and reformed in 1966. *Plus de cargas familiares* was repealed and the *subsidios familiares* was updated. According to the 1966 reform, family benefits were granted to all insured workers for dependent children and dependent spouses. The amount of the child benefit was made uniform and independent of family size.

In addition to the child allowance, family benefits also included a monthly allowance for an economically dependent wife or disabled husband. This benefit was fixed at 300 PTAs (8.3 per cent of SMI in 1970) until 1971, then raised to 375 PTAs in 1975 (4 per cent of SMI) until 1985 (when it was 0.9 per cent of SMI).

In addition, the family benefits system granted a marriage allowance of 5,000 PTAs from 1966 to 1971. Families also received an allowance for childbirth worth 2,500 PTAs from 1966 to 1971, which was then raised to 3,000 PTAs until 1990. These allowances, directed at the nuclear family, marked a change from past legislation. There was a shift from conceptualizing the family and its legitimate dependent members in a broad, extended sense to limiting the definition to a couple and its children. As far as the amount of the allowance is concerned, its real value was frozen in 1971 and subsequent high inflation rates nullified its value within a few years. The constant devaluation of family allowances in the Spanish social security system can also be seen in the data provided by Meil (1994). The percentage of social expenditure for the family (family, marriage and childbirth allowances) decreased from 2.6 per cent of GDP in 1965 to 1.5 per cent in

1973 to 0.8 per cent in 1976.

Child care for the youngest and educational policy During the Franco regime, most day-care services and education services for very young children had been provided by the private sector, with the Catholic Church accounting for a large proportion of services. In addition, the Catholic Church controlled the Ministry of Education, which it used to indoctrinate Spanish youth. In 1945 a Primary Education Act was passed which differentiated day-care centres for children under the age of 4 (*escuelas maternales*) and pre-primary schools for children from 4 to 6 years (*escuelas de párvulos*). These services were only intended for very young children of poor working mothers and thus their purpose was more auxiliary and custodial than educational.

In 1970 a General Law of Education[13] increased the duration of compulsory schooling from 7–12 years to 6–14 years. This law also gave the state responsibility for organizing child care. The law established two levels of pre-primary education: nursery schools (*jardín de infancia*) for children aged 2 to 4, and kindergartens (*párvulario*) for children aged 4 and 5. Despite this law, as many authors have recognized (for example Palacios, 1989; Berea, 1992), pre-school education was not a real priority of the 1970 law. In fact, the number of places in pre-school classes actually decreased. It can be said that at the end of the Franco regime, pre-primary education in Spain was still underdeveloped and there was a general lack of child care facilities, especially in the public sector.

Family Law Under Franco

At the end of the civil war (in 1939), Franco abolished all the legislative innovation introduced during the Second Republic, including women's right to vote, divorce by mutual consent, equal rights between spouses, and abortion. Franco did not follow this annulment of progressive legislation with a new family law and/or new definitions of family relations, but rather reactivated the traditional family law of the 1889 civil code. This, as I have shown, was largely based on a profoundly unequal distribution of power between genders and across generations within the family. Thus, civil law under Franco starkly confirmed women's total subordination to marital authority and all the principles related to *patria potestad*. Legally, women were totally dependent upon their husbands. The *Fuero de el Trabajo* proclaimed the achievement of the 'freedom' of women from work, while it actually forbade married women to do so. Women's access to most public professions was forbidden: a woman could no longer be a notary, a diplomat, and/or employed in the public administration. Marital authorization was required for any application to work (Bautista Parejo, 1996).

The only legal changes during the first years of Franco occurred in abor-

Table 4
SPAIN, DEVELOPMENT OF THE WELFARE STATE AND STATE POLICY TOWARDS THE FAMILY

	The take-off period (1900–30)
1900	*Maternity compulsory leave period (3 weeks)
1907	*Maternity compulsory leave period reformed (4 weeks)
1908	Creation of National Institute of Insurance (aimed at promoting social insurance among low-income workers)
1919	Old-age insurance
1923	*Maternity benefits
	The second republic (1931–36)
1931	*Maternity insurance scheme reformed
1932	Occupational injury insurance
1931–36	Statement by the Republic on Constitution that it was the responsibility of the state to create a system of social insurance against illness, labour accidents, unemployment, old age, disability and maternity. However, legislation was submitted to parliament just before the outbreak of the civil war.
	Civil war (1936–39)
1938	**Subsidio familiar* (family allowances)
	Franco regime 1939–59: period of autarchy
1939	Old-age subsidy for low-income workers over 65 years or for those with disabilities over 60 years – consolidated in 1947 as the Old Age and Invalidity Insurance Programme
1942	*Compulsory health insurance for low-income workers and their dependants (coverage 30% of population in 1946, rising to 44% by 1960)
mid-1940s	*Mutual Aid Associations (providing benefits for retirement, disability, long illness, widows and orphans and hence duplicating the insurance programmes)
1945	**Plus de carga familiares* (family bonuses)
1954	*Family bonuses for civil servants
1955	*Survivors' pension
	Franco regime 1959–75: the capitalist period
1963	Basic Law of Social Security
1966	*Unification of the family allowances system
1972	Financing Law of Social Security

Note: Law and social policy related directly to family are marked by an asterisk.

tion and marriage law. The abortion law of 1941 established a link between population increase and abortion. Abortion and the use of contraceptives were no longer considered a crime against human life, but a crime against the state, whose duty it was to preserve the race. The marriage law of 1938 eliminated the non-confessional matrimonial regime enacted in 1932 and reintroduced the influence of Catholicism and canon law on Spanish marriage law until the 1978 Constitution. Afterwards, as a result of the Concordat between the Church and state in 1953, the canonical marriage in Spain had a civil effect and only formally declared non-Catholics were allowed to marry in civil ceremonies (Alberdi, 1994).

However, there were some legislative changes during the decades before democracy that aimed to reduce the inequality between spouses and to change women's traditional role in society. In 1961 a law relating to women's political, professional and employment rights and against sex discrimination was introduced. This law gave women access to the public professions and prohibited their dismissal on grounds of marriage. In 1966 Spanish women could become judges. In 1975, Spain abolished Article 57 of the civil code, which stated that a husband must protect his wife and a wife must obey her husband.

NOTES

1. Prior to the EPA, census data underestimated the number of women working in the agricultural sector and in home work. On the question of using historical data series statistics for Spain see Carreras (1989) and López (1988).
2. On the issue of women's employment as a source of conflict within the working class, see the text of Virgina Gonzalez (who was a primary actor in the Communist Party): 'A las obreras', text no. 72, p. 310, in Nash (1983).
3. These self-labelled Spanish 'neo-liberals' did not have much in common with the term used to describe the turn taken in economic policy in the late twentieth century.
4. Law no. 88, 13 March 1900.
5. Law, 1 January 1907.
6. Real Decreto (RD), 11 March 1919.
7. RD, 21 August 1923.
8. The project was approved in 1929 (RD, 22 March 1929), the regulation passed in 1930 (RD, 29 January 1929).
9. On the issues of social policy during the dictatorship of Primo Rivera and during the Second Republic, see Cuesta Bustillo (1988) and Samaniego Boneu (1988).
10. Law decree of 26 May 1931. The law was implemented on 1 October 1931.
11. For an analysis of the nature of Francoism, see Giner and Pérez Yruela (1979) and Pérez Yruela and Giner (1988).
12. Law, 14 February 1942.
13. Law no. 4, 4 August 1970.

PART THREE

THE DEMOCRATIC PERIOD: A FAMILY/KINSHIP SOLIDARITY MODEL

INTRODUCTION: THE WEAKENING OF THE MALE BREADWINNER

In the second part of this book analysis of the origins of state policies towards the family has illustrated that, at their height, the two dictatorial regimes supported a definition of the family in which several features of the male breadwinner family model were embedded in social policies and laws. In general, social policies and laws were aimed at supporting the new figure of the adult male working in the urban setting, free from the rural patriarchal and kin-oriented family model. However, these two countries hardly achieved this family model in its entirety. As an 'ideal type', the male breadwinner model requires 'full male employment', high wages, widespread industrialization, and the prevalence of a bourgeois family lifestyle situated in an urban setting.

There are a number of reasons why Italy and Spain only partially achieved the male breadwinner family model. First of all, compared to other European countries, the different socio-economic contexts in Italy and Spain limited the degree to which these two states could completely support the male breadwinner model. Italian and Spanish occupational structures were marked by regionally uneven modernization and a belated decline of the agricultural sector. Traditional rural society was also preserved through small enterprises that developed within the agricultural settings. The long-lasting viability of rural life in Spain and Italy had important consequences for the male breadwinner model. Most people still lived in small communities, retaining rural and traditional family systems and values, and yielding to the moral regulation of local Catholic parishes. In this context intergenerational relations and kin solidarity were expected and perceived as having strong moral foundations. Not to forget that in both societies a high number of women had to work for pay in the informal labour market and/or in the agricultural sector. Taken together, these economic, social and cultural factors could not permit a complete achievement of the male breadwinner family model.

The permanence of 'traditional' family and kin systems and rural family values in Italy and Spain shaped not only the support for but also the assumptions and principles underlying the family/kinship solidarity model. Italian and Spanish fascist-era social policies aimed to support the breadwinner and took into account not only the nuclear and 'urbanized' model of the family, but also the 'rural' and traditional family based on intergenerational and gender obligations and kinship solidarity. Therefore, the states' concerns had to be not only the nuclear family, but also kin. State policies towards families, through the figure of the male breadwinner, made it possible to funnel social protection not only to dependent wives and children, but also to parents, orphaned siblings and unmarried sisters. Translated in terms of family models, the Italian and Spanish welfare states during their dictatorial periods have been characterized by a sort of 'stretching' of the male breadwinner family model when the social and cultural context was favourable to the family/kinship solidarity model.

The third part of this book is devoted to showing what happened in these two countries in the democratic period, following the fall of the two fascist regimes. In particular, it aims to illustrate how and if the new democratic governments in Italy (since 1948) and Spain (since 1975) maintained or dismantled the unique family/kinship solidarity models which had been embedded in the previous era's policies and laws.

The central proposition developed in this part of the book is that the definitions of the family as found in laws and social policies during the democratic Italian and Spanish welfare states, have maintained most of the past characteristics of the male breadwinner family model that stretch towards the family/kinship solidarity model. The next chapters will illustrate that while most of the characteristics identifying the male breadwinner model are sometimes still found, it is not because of any active policy-making explicitly promoting that model. Instead, these characteristics are the remains of legislation of the previous eras.

The most relevant change in comparison with the dictatorial regimes has to be sought in the removal of pro-natalist family policies. The repeal of pro-natalist social policies has resulted in a progressive withdrawal of state support to fathers for the cost of children. This withdrawal of state support has weakened the already feeble male breadwinner family model in the Italian and Spanish historical cases.

In the next four chapters the evolution of state policies towards the family will be analysed, and the main factors which account for the explanation of the weakening of the male breadwinner and the persistence of the family/kinship solidarity model will be explored, taking into account the wider political, legal and sociological contexts that characterized these societies.

Chapter 5, 'The Political and Legal Context', illustrates the political transition towards democracy and the main features of the Italian and

Spanish political party systems. Next it explores the place given to the family in the Italian and Spanish Constitutions, and the political and ideological divisions regarding family matters in the two historical cases, as well as the main political orientations on the family. This will provide a few explanatory variables to help to account for the current institutional definition of the family model. The analysis of political context demonstrates, especially in the Spanish case, that the identification of the 'family' and of 'family policy' with authoritarian pro-natalist and anti-feminist policies prevented any intervention in the area of social policies for family responsibilities.

The analysis of the political context will also show to what extent the Italian post-war configurations and in particular the ideological division between communists and Catholics in Italy have contributed to the weakening of the male breadwinner family model and to the strengthening of the family/kinship one. In Italy 'family matter' became a source of tension between Catholics and communists during the drafting of the Constitution. Since then, and for a long time after, neither the Christian Democrat Party nor the Communist Party had an interest in openly facing Italian family matters. Finally, the last part of Chapter 5 provides a picture of the way in which gender, generation and kin relations are defined by Spanish and Italian civil law. This part of the analysis is important not so much because it shows the persistence in Italy (up to the mid-1970s) of a traditional and patriarchal definition of family relations, but insofar as it shows the persistence in both Italy and Spain up to now of a strong definition of family and kin obligations and the strong influence of the Church in divorce law.

Chapter 6, 'Family Changes and Family Solidarity', explores family changes in Italy and Spain from 1960 to the 1990s within the context of socio-economic development, family law reforms and the expansion of the welfare states. This chapter shows that despite the presence of profound changes in family behaviour during recent decades, these countries are still characterized by traditional forms of matrimonial and intergenerational ties, as well as by a shared cultural family model in which kin solidarity is still expected. The persistence of the family/kinship solidarity model will also be related to a very specific feature of the functioning of the 'dualistic' protection system in Italy and Spain in discussing modernization processes and welfare state expansion.

Chapter 7 illustrates how social policies and social services for families have changed during the two democratic periods. The focus is placed on four policies: survivors' pensions; family allowances; tax relief for dependent family members; and maternity and child care provisions. The main goals of this chapter are to find out which definitions of family models stem from which laws and policies, and to discuss to what extent the definitions have changed from those of the authoritarian periods. Comparing social policies and social services as a whole, it will appear evident that notwithstanding their constitutional premises, neither the Italian nor the Spanish welfare state developed any form of 'family policy' and/or important social policy aimed at socially and economically protecting families with children. Instead, the two democratic welfare states simply broke with fascist-era, pro-natalist policies while leaving the overall institutional structures of family policy, especially family allowances, intact.

The Italian and Spanish welfare states during the democratic period have been characterized not only by a lack of any explicit family policies, but also and especially by a systematic decline over time in the level of public transfers to families with children. For a long time, social policies related to family responsibilities, and to the cost of children, have consisted only of meagre tax credits and a reduced number of social security and contributory benefits, which are available only to dependent workers. Especially during recent decades, state support for families with children has declined. Public economic support for families with children has been very low, not only in comparison with the previous period but also in comparison with other European countries.

Finally, Chapter 8 presents an in-depth study of policy-making for four policy areas. It will illustrate that, in both countries, during the democratic period the issue of the costs of children was never fully tackled. The analysis of this issue is important because it helps to explain why in both countries a real reform of family allowances is long overdue; why the level of public support for families with children is so low in Italy and Spain, both from a synchronic and diachronic perspective. Finally, it helps to explain why caring for children in the Italian case, especially the very young, is still perceived as a private, family (i.e., woman's) responsibility.

Chapter 5

THE POLITICAL AND LEGAL CONTEXT

The Political Configuration

Transition to Democracy

There was a difference of 30 years between the democratic transition and subsequent promulgation of the Constitutions in Italy and in Spain. Thus, the process of democratization occurred in significantly different geopolitical, international and socio-economic contexts in the two countries. This makes historical comparison of the political and legal context somewhat difficult. In Italy the process of democratization began after the end of the Second World War, a period which was characterized by sharp internal divisions. As a matter of fact, from 1943 to 1945 the country was split into different polities. The end of the war and the defeat of fascism, which was marked by the period of liberation, resulted in the first free elections in 1946, and the right of women to vote for the first time in Italian history. The transition to democracy was marked by the formation of a strong alliance between all the main anti-fascist forces and resulted in the promulgation of the Constitution in December 1947, which was implemented on 1 January 1948.

In Spain the transition to democracy began with the death of Franco in November 1975. In June 1977 the first democratic elections since the Spanish Civil War took place. The democratic transition involved the principal government and opposition forces and was strongly supported by trade unions (the Unión General de Trabajadores, UGT, and the Comisiones Obreras, CCOO) and employers' organisations (the Confederación Española de Organizaciones Empresariales, CEOE), as is evident in the Moncloa Pacts.[1] These pacts established a policy of mutual restraint between the government, the employers and trade unions, with the aim of 'consolidating democracy'. The first phase of transition ended on 6 December 1978 with the promulgation of the Spanish Constitution. The next sections are an attempt to understand if the 30-year gap between Italy's and Spain's democratization has also been relevant to political and legal definitions of the family.

Party Systems and the Context of Competition

The Italian party system After the end of the Second World War the most obvious sign of the resurgence of democracy was the refounding of mass political parties: the Communist Party (Partito Comunista Italiano, PCI), the Socialist Party (Partito Socialista Italiano, PSI) and the Christian Democrat Party (Democrazia Cristiana, DC), heir of the Partito Popolare (PP). The two largest and strongest parties, the Christian Democrats and the Communists, corresponded to the largest and most deeply rooted political traditions in the country: Catholic and socialist/communist. Unlike other democracies, after 1946 the party alignment in Italy was not clearly divided between the left and the right. Rather, party alignment was split three ways, divided into centre, right and left parties. The presence of a large, profascist extreme right prevented the formation of a large 'constitutional' right and favoured the formation of a large centre party (Farneti, 1985). At the same time, the presence of a large, pro-Soviet extreme left prevented the formation of a large 'social democratic' left, thus also encouraging the formation of a large centre party. This three-way alignment of right, centre and left gave other smaller parties – the Republicans (Partito Repubblicano Italiano, PRI), the Liberals (Partito Liberale Italiano, PLI), the Social Democrats (Partito Socialdemocratico Italiano, PSDI) and later the PSI – the opportunity to position themselves both between the centre-right and centre-left. The DC, which was the party of relative majority for more than 40 years, located itself at the centre, rather than at the right, and until the beginning of the 1990s was the main governmental party in collaboration with other parties that refused to co-operate with the extreme right party, Movimento Sociale Italiano (MSI) and with the PCI (Farneti, 1985).

Two types of governmental coalitions characterized Italian post-war politics up to the 1990s. The first were 'centrist coalitions' led by the DC together with smaller parties: the PLI, the PRT and the PSDI. The first democratic government (1947) excluded both the Socialist Party and the Communist Party, bringing about the political period known as 'centrist government', which ruled over Italian politics from 1948 to 1963. The expulsion of the left (but especially of the Communist Party) from the first republican government has to be seen in light of the deep ideological division that has characterized Italian politics since the beginning of the Cold War. Because Italy was under strong American influence and the PCI was considered an ally of the Soviet Union, the Italian Communist Party remained excluded from the government for a very long time.

The centre-left coalition (DC–PRI–PSDI–PSI) was the second type of coalition that characterized Italian politics. This coalition came together in 1963, when the Socialist Party, formerly the opposition party, went into government with the DC. This coalition lasted in parliament first for twelve years (1964–76), and then with a revival in the 1980s that lasted until the beginning of the 1990s. The centre–left coalition was interrupted by a three-year interval between 1976 and 1979, when the Communist Party pursued its strategy of 'historic compromise' and externally supported the three national unity governments led by the DC leader, Giulio Andreotti.[2]

From 1980 until 1993 Italian politics was dominated by a coalition of the DC and the PSI (in collaboration with PSDI, PRI and PLI, the so-called *pentapartito*). The PSI played a pivotal role, and its leader, Bettino Craxi, was Prime Minister for a very long period (1983–87) by Italian standards. The 1980s were years of great historic continuity with the previous centre-left period; they also fully demonstrated the limited capacity for reform in a 'blocked' Italian political system. According to Pasquino (1996), one of the main features which distinguishes the Italian political system from those of other democracies is the supremacy political parties held over cultural, political and social associations in civil society. This supremacy was developed through a penetration of political parties, especially the DC, into all aspects of society that had begun with the democratic period. The only associations outside the dominion of political parties were those supported by the Catholic Church.

As far as policy outcomes are concerned, participation in highly fragmented and ideologically divided coalitions made it difficult for individual parties to promote and then implement their own social policies. As a result, minor laws for selected problems and clients have proliferated in the Italian welfare state. The case of pensions is a well-known example. The coalition style of government has also hampered all fundamental reforms involving structural changes and/or addressing very conflict-ridden issues (for instance divorce law).

Between the end of the 1980s and the beginning of the 1990s the Italian party system saw profound transformations, which also changed the traditional coalitions. During the 1994 election campaign the traditional and largest parties, such as the PCI, DC, PSI, and the MSI, split or redefined themselves and totally new parties – the Lega-Nord and Forza Italia – emerged. As a result, new coalitions overturned traditional alliances. This was particularly true within the centre-right: the newly dominant Forza Italia, made up of former socialists, liberals, Christian democrats and individuals who had never before engaged actively in politics, allied itself with the post-fascist Movimento Sociale Italiano/Alleanza Nazionale (MSI/AN), the right wing of the former Christian Democrats (Centro Cristiani Democratici), and the Radical Party. This right-wing coalition was in government for only one year, when it was overturned by 'Ulivo', a new centre-left coalition that emerged for the election campaign of 1995. This coalition was made up of the PDS, the left wing of the former Christian Democrats (Partito Popolare Italiano, PPI), and received the external support of Rifondazione Comunista.

The Spanish party system The contemporary Spanish party system is very young, since it came into being only a few months prior to the June 1977 general election. Since then it has been characterized by extreme instability. This has been manifest in the changing number of parties, their variable electoral strength, and the substantial weakening of the party–electoral bond. Consequently, even now it is extremely difficult to find a precise definition of the Spanish party system. According to Cotarelo (1992) the Spanish party system is characterized by a high degree of political moderation, a high degree of fragmentation (augmented by the presence of regional parties) and ideological polarization. The Spanish party system has been defined in various ways, with reference to the number of competing parties (fragmentation), and/or to the ideological distance (polarization) between them. It has been characterized as 'two-party', 'imperfect two-party', 'two-party-plus', 'polarized pluralism', 'segmented pluralism', 'fluctuating party', 'predominant party', 'limited multi-party with tendency towards bipolarity', and non-crystallized, non-stabilized' (Heywood, 1995: 174). The difficulty in classifying the Spanish party system is, according to Cotarelo (1992), related to two factors. Firstly, the difficulty in creating a right-wing party. Secondly, the existence of a regional sub-system of parties, which were important at the national level. The Spanish party system not only lacks consolidation, but is also still evolving. In order to understand the nature of party politics in Spain, we must devote some attention to the history, aims and internal functioning of the parties, as well as to the political context in which they operate.

During nearly 40 years of dictatorship under General Franco, political party activity was illegal. This had a significant impact on the political parties that emerged during the post-Franco democracy. Only a few parties participating in the first election could claim continuity with those that existed prior to the dictatorship. The Communist Party, Partido Comunista de España (PCE), and the Socialist Party, Partido Socialista Obrero Epanañol (PSOE), were two of the few who could make this claim. They were also the only two parties that contributed to the downfall of the regime. Even then, however, both parties had experienced changes both in their ideological content and political strategy. All right-wing post-Franco parties were of recent creation, including both the Union de Centro Democratico (UCD) of Adolfo Suarez and the Alianza Popular (AP) of Manuel Fraga.[3]

In the first 1977 general election only five national level and two regional level political parties got representation in the Spanish parliament. At the national level, the UCD and the PSOE were the main political forces, receiving 34.8 per cent and 29.9 per cent of the votes, respectively. The PCE, which stands left of the PSOE, received 9.3 per cent of the votes, while the party right of the UCD, the AP, received 8.4 per cent. At the regional level, Convergencia y Unió (CiU) (3.7 per cent) and Partido Nacionalista Vasco (PNV) (1.7 per cent) got representation in parliament. The 1979 general election more or less confirmed the prior results. According to Heywood (1995) the Spanish system during the first period from 1977 to 1982 (with UCD being the majority party and the main governmental force, and the PSOE the main opposition party) can be defined at the national level as an 'imperfect two-party system' or 'moderate pluralism'.

The 1982 election marked clear changes in both the party system and in the political force in power. In 1982 the UCD (the centre party) virtually disappeared, and the PSOE won the election, obtaining an absolute majority with 47.3 per cent of the votes. Following this election the PSOE was the main political and governmental force until 1993. According to Heywood, the period 1982–93 (with the PSOE being the party having the absolute majority, and Alianza Popular/Partido Popular (AP/PP) the main opposition force) can be defined as a 'predominant party' system. He also argues (1995) that the Spanish political party system has experienced a decline of ideological distance between parties. In other words, polarization within the Spanish party system, especially since 1982, has been much lower than in other European countries.

The Place of the Family in the Constitution

The Italian Constitution

The work of the Italian Constituent Assembly lasted 18 months and resulted in a democratic constitution which unified all political parties within a common matrix of anti-fascism. The Italian Constitution makes declarations on social rights and gives the family an institutional place in society. This constitutional recognition of the family is not peculiar to Italy: Germany, France, Greece, Ireland and others also recognize the family in their constitutions. Importantly, constitutional acknowledgement of the family has not been the same across all countries. For example, the Portuguese Constitution declares that the family, as a fundamental institution of society, needs to be adequately protected. Others, like Germany's, emphasize protecting marriage (CNAF, 1995).

The Italian Constitution was the result of a set of historical compromises. On matters relating to the family, the forms of the constitutional compromises are fixed in some articles. First of all, the constitutional compromises have been embedded in Article 7, by which the Concordat between the Church and the state of 1929 (the so-called 'Lateran Treaties') became part of the Italian Constitution (with the approval of the Communists). This article established a two-fold privileged status for the Church in the new republic. The first is concerned with the matrimonial clauses of the Concordat that attributed civil effects to marriage celebrated according to canonic law. The second concerned the position of religious education in state schools.

Besides the Concordat issue, the nature and the role of women and of the family were central to the debates in the Constituent Assembly and were often sources of division among the main political parties. Five fundamental questions were raised: the definition of family; equality between spouses and the representation of women's roles; the status of children born out of wedlock; the indissolubility of marriage; and the degree of autonomy of the family *vis-à-vis* the state.

1. As far as the definition of family is concerned, the Constitution recognizes the family as a 'natural society' based on marriage (Article 29).
2. The relationship between sexes within marriage and women's role within the family and in paid work are mentioned in the second paragraph of Article 29: 'marriage is based on the moral and legal equality between the spouses, within the limits of the law protecting family unity'. The principle of equality between genders within the family is placed in contrast to two other principles of unity and solidarity. Men and women are equal, except where the unity of the family is concerned (Saraceno, 1998; Vincenzi Amato, 1988). In Article 37 the Constitution explicitly establishes that equality between the spouses is subordinate to the fact that woman's employment outside the home should not interfere with the performance of a woman's essential role in the family, or *l'adempimento della sua essenziale funzione familiare*. The constitutional representation of women's role is primarily as mother and wife and paid work is subordinated to this.
3. The issue of the status of children born outside of marriage is stipulated in Article 30: children born within and outside of marriage are considered equal, within the limits established by the law in order to protect the legitimate family.
4. The indissolubility of marriage, regarded as a principle to be declared in the Constitution, failed by a few votes. Divorce was not assumed in the legal system.

5. The issue of the relationship between state and family, chiefly, the division of responsibility in the areas of reproduction and caring, was an untouchable issue. The Italian Constitution stipulates that the family has to be adequately protected, and that the state provides the protection of maternity, childhood and youth (Article 31, paragraph 2) through family allowances and other provisions, in order to make it possible for the family to accomplish its functions (Article 31, paragraph 1). Notwithstanding these constitutional premises, most constituents were opposed to state intervention in family matters and the government never planned any form of family policy and/or social policy aimed at the family.

The Constitution's approach to family relations and the role of women seemed more inspired by the traditional, Catholic authoritarian view of the family than by the new ideals of democracy and of juridical and political gender equality found in other parts of the Constitution.

During the 1950s and the first half of the 1960s the tension between the principles contained in the Italian Constitution allowed Italian judges to adhere more often to the principles of the laws protecting family unity and the legitimate family, rather than those of gender equality. Although there were hints of change prior to 1975, it was not until the reform of family law in that year that husbands and wives were no longer treated differently under the law. The same held true for children born within and out of wedlock (Vincenzi Amato, 1988; Saraceno, 1998).

The Spanish Constitution

The 1978 Spanish Constitution established a new political regime. It took the form of a parliamentary monarchy, based on the principles of freedom, democracy and political and religious pluralism. In particular, it is important to remember that the political transition involved a change from a 'religious dictatorship' to a democratic and pluralistic juridical and political system. The Constitution reflects such political transformations in the principles it expresses in the fixed articles concerning the family. The Spanish Constitution affirms three main principles: equality between men and women; the possibility of dissolution of marriage; and the principle of protecting the family and equal protection of children regardless of their legitimacy status. These principles are contained in Articles 32 and 39.

1. First, the interest in legally fixing the political and cultural change led to the declaration in Article 32 that 'both man and woman have the right to contract matrimony with full legal equality'.
2. Article 32 also appealed for a new law regulating the grounds and con-

sequences for dissolution of a marriage: it states that 'the law will regulate the forms of marriage, the age and the capacity to contract it, the right and the duties of the spouses, the causes of separation and dissolution and their effects'.

3. Finally, we see how the Constitution supports the principle of protecting the family. Article 39 states 'public power ensures social, economic and legal protection to the family'. The Constitution also establishes the basis of the state's responsibilities in the protection of children and for social and economic policy within this sphere of social relationships. It states 'public powers ensure the total protection of children, who are equal in the law, regardless of their consanguinity, their mothers, whatever their marital status.'

Despite these principles, during the democratic period Spain did not develop any forms of social and economic support for families with children. The next section attempts to explain this by looking at the political orientations of the main political and social actors. Unlike its defining approach to marriage and progeny, the constitutional text leaves the definition of family relations to the future civil code (1981). This allowed lawyers to adapt legal provisions to new and changing family situations and to leave room for future social evolution (Alberdi, 1992).

The main constitutional principles regarding the family were incorporated into the civil code, which was reformed in 1981. However, the principle of social protection of the family, which would be accomplished through state policies and the state provision of services, was not implemented in the welfare state reforms.

The Family as a Political Issue

How did the ideological and political configurations in Italy after the Second World War influence the institutional definition of the family model? To what extent did the tensions of the constitutional compromise on family matters, and the deep and long-lasting ideological division between communists and Catholics, contribute to influencing family law policy-making and, campaigns. The centre-right parties in Spain constitute the only exception: they have attributed high institutional value to the family and supported social policies for family responsibilities since their first manifesto.

The Italian and Spanish political parties' omission of the family and family issues from their political platforms should not be taken for a lack of concern. On the contrary, parties in both countries approached family issues within other contexts, policy-making on employment, women's policy, and the legal rights of individuals. In the two national cases under consideration, political parties and other social actors intentionally avoided addressing family matters and this may be because the family, its values and its symbols were too sensitive for political parties to handle.

Analysis of Italian and Spanish political party platforms reveals many similarities and some differences between the two national cases. A similarity can be found between the Italian and Spanish left-oriented political parties. In both countries, these parties long avoided confrontation on family matters and defining what the family is and what family model should be supported.

Moreover, in both Italy and Spain any social policy demand for financially supporting families with children (i.e. family allowances) was – in political party platforms – systematically avoided, delayed or passed to other social actors (i.e., to trade unions). It should be kept in mind that in both countries family allowances were primarily a social security benefit, and thus considered a labour policy rather than a social policy for families.

More recently, analysis of the orientations of the main political actors shows that in both countries political party programmes gave new attention to family issues between the end of the 1980s and the beginning of the 1990s. This new attention was partly the result of the welfare state crisis, and, I would add, partly a result of renewed public concern about fertility decline. The trend in decentralizing political and welfare policies also contributed to increased demands for family policies. Several regions in Italy approved laws on family policies. In addition, some local governments at the municipal level developed explicit sets of policies addressing family responsibilities, either in the area of child care or of the elderly. The same occurred in Catalonia in 1993, when the nationalist political party (Convergencia i Unió) approved a plan for supporting families. Spain also experienced some decentralization of social policies and services related to child care, the elderly and low-income families. This process of decentralization of social policies for families highlights two factors. First, within each country different family cultures are observable. Second, in countries so deeply divided on the definition of 'the family', the likelihood of producing innovations in social policies not for 'the family' but for different family forms, is higher when the decision is taken at subnational levels (i.e. at *Regione, Comunidad Autonoma, and Comune*).

Despite these similarities, there are also important differences between the two national cases. A fundamental difference between the Italian and Spanish historical cases is linked to the important role (in raising social demands) played in Italy by social movements between the end of the 1960s and the mid-1970s. Another important difference was the attitudes of the feminist movements. While the Italian feminist movement struggled actively not only for women's rights and individual rights, but also for public child care, Spanish feminists started to demand public child care policies only in the mid-1990s.

The Long Absence of the Family in Italian Politics

During the drafting of the Italian Constitution family issues had incited tension and conflicts between the main political forces (Bimbi, 1993). As I have shown, the solution (although ambiguous and partial) was found in a set of constitutional compromises. For a long time after, Catholics and Communists, the two main political actors and ideologies controlling the political arena and public discourse up to the mid-1960s, had little interest in reopening the family matter (divorce law was introduced in Italy only in the 1970s), or in advocating any specific family policy (Saraceno, 1998). In other words, in Italy following the approval of the Constitution, family issues were no longer an explicit target of intervention in any party platform and in political discourses until the late 1960s.

Since the drafting of the Constitution, Christian Democrats have defended the idea that the family has to be regarded as a social unit based on natural law. This standpoint was inspired by Catholic doctrine, which the DC used as a basis for the moral, legal and social regulation of family relations and for defining women's role in society. According to this view, which was also embedded in the Italian Constitution, women's role in the paid labour market is subordinate to their role as mothers and as wives. In October 1945, Pope Pious XII had declared his opposition both to the principle of gender equality and to the practice of women working outside the home, proclaiming that 'woman's destiny is to be mother' (quoted, in Ballestrero, 1979: 114). Moreover, according to the principle of 'subsidiarity', in Catholic social doctrine the family is viewed as a private, autonomous and self-sufficient world capable of meeting the needs of its members. The state should interfere as little as possible, apart from offering support in order to guarantee family unity and stability, and apart from upholding the gender division of labour and responsibility.

In terms of public policy during the 1950s and the 1960s, the Christian Democrat Party supported those social policies that strengthened the role of the male breadwinner. In fact, DC members conceptualized family allowances as supplementary to the husband's wage insofar as male wages were not always adequate to support a family (Falcucci, 1967). Accordingly, the DC also supported those policies which aimed to reinforce the social value of motherhood (for instance, long maternity leave) and the role of women as wives.

Indeed, in the Italian political and ideological context there was no disagreement between the Catholic and fascist visions of family relations and of women's role in society. The Church disagreed, instead, with the dictatorial regime's imperialist and racist policies. Bearing in mind that fascist and Catholic ideologies of the family were very close, it is not surprising to observe that the DC did not ask for any family law reform in the new republic. The DC's lack of advocacy for family policy can be explained by two further factors. First, there still exists a profound reaction against the pronatalist family policy developed by the fascist regime. Second, any further intervention in family matters was seen as encroaching on the territory of the Church. Catholics, however, were not against all kinds of child care services and/or services for families. Their opposition was towards those that were not organized by the Church and/or Catholic associations (Saraceno, 1998). The Christian Democrat Party recognized the possibility of state-run child care services only when the family had exhausted its own resources, that is, in cases of need. Services could be provided, for instance, in the case of poor working-class families in which women were obliged to work for pay.

On the other hand, the Church in Italy, up to the middle of the 1960s, remained the 'natural' agency for providing educational services for children and social assistance services for the elderly, and it was the first recipient of state subsidies. The state limited its intervention policy to assistance for the extremely poor (Fargion, 1998).

The Communist Party position on family matters was no less ambivalent than that of the Christian Democrats. Throughout the period between the approval of the Italian Constitution and the end of the 1960s, the PCI avoided elaborating its own position on family relations and on social policies for families, and avoided declaring it publicly. The PCI was very careful not to break with the Italian Catholic tradition and with the 'religious feelings of the people', particularly those of women. This party had maintained a 'pro-family' line and was careful not to be perceived as the 'pro-divorce' party (*il partito dei divorzisti*). It had to keep a position that would satisfy its patriarchal working-class constituency without alienating the Catholic Church. The PCI leader Palmiro Togliatti encouraged communist women to co-operate with Catholic women on family issues, and criticized those 'women companions' (i.e., Unione Donne Italiane (UDI) women), who accused Catholic women's organizations of opposing any attempts at women's emancipation (Togliatti, 1965; Caldwell, 1991; Bimbi, 1978). In 1945, during the first Conference of Communist Women, Togliatti advocated the need to maintain family unity. He argued that in a deeply unsettled country, as Italy was right at the close of the Second World War, there existed the need to rebuild and defend family unity. Finally, he argued that the family had to become a centre of elementary human solidarity (Mancina, 1981). Twenty years later, when the fourth Conference of Communist Women proposed the first family law reform, and within it divorce, Togliatti replied that Italy was not yet ready to introduce divorce (Gramaglia, 1974).

In terms of public policy the PCI (as well as the PSI) supported both cash-related family benefits for male workers and social rights for women, provided that these were granted in support of all workers' rights or in the name of women's rights as paid workers (Bimbi, 1992). Therefore, their position in this field was not far removed from that of the Christian Democrats.

The field in which the Communist Party was in conflict with the Catholics was women's emancipation and rights. The PCI, inspired by the Marxist theory of women's emancipation through employment, supported, along with trade unions and the left-leaning PSI, equal rights for women. Especially important were equal economic rights and the rights of working mothers. So, on family issues the Communist Party preferred to restrict its political focus to the rights of working women, while implicitly denying that women's participation in the paid labour force posed any problems either for the traditional organization of the family or for women themselves (Mancina, 1981).

Family issues and legal changes during the 1970s Things began to change in the mid-1960s, when, as a result of welfare state expansion and new demands from social movements, new family issues were raised in society. At the end of the 1960s the issues of working women's rights and child care were put on the political agenda. In this period kindergartens (1968), publicly funded day-care services (1971) and the new law on maternity leave (1971) were introduced. Thus, political parties were obliged in some ways to reopen the family debate they had closed 20 years before. Pressure from the feminist movement and the main 'lay and liberal' political forces (represented by the PSI, PRI, PLI and PR) also pushed family issues onto the political agenda through the mid-1970s. But at this time the debate on family issues had shifted from child care and women's emancipation through work to the issue of legal changes. Since the end of the 1960s family law reforms attained particular importance within the political and parliamentary debates and within society, especially as the issue became one of the most important claimed rights of spontaneous movements (youth associations, feminist movements, extra-parliamentary movements, and so on). The debate, within the legal framework of family relations, focused on the redefinition of gender and generational hierarchies. By 1967 Italy had already approved its first adoption laws. In addition, it has to be noted that jurisprudence during the second half of the 1960s had already begun to reflect important changes in values and behaviour and that the courts had issued decisions more in line with the equality principles of the Constitution (Pocar and Ronfani, 1995).

In 1970 the divorce law was introduced, and in 1975 family law reforms[5] profoundly changed the hierarchical division of power between genders and across generations within the family. In the same year the age of majority was lowered to 18. Equal treatment between men and women with respect to paid work was introduced in 1977. The use and the advertisement of contraception were legalized in 1971, and family planning clinics were introduced in 1975. Finally, abortion was legally authorized in 1978 (the referendum was in 1981). These family law reforms profoundly altered the 'traditional' structure of the family inherited from fascism, Catholic ideology, and a shared working-class and peasant milieu (Del Re, 1995; Bimbi, 1993; Pocar and Ronfani, 1995).

It can be noted that it was during the electoral campaign of 1968, and as a response to what was going on in society, that two important Italian political parties began to address several family issues. In 1968, the Christian Democrat Party electoral platform contained two issues addressed explicitly to the family. First, they proposed to support the husband's wage through an increase in family allowances, and to support the role of women as mothers through a long maternity leave (DC, 1968). The second proposal was related to legal changes (family law reform). The DC advocated equality between the spouses in the field of parental authority, but opposed divorce in the name of preserving family unity.

The Socialist Party was the other political force that addressed family issues in its electoral platform. The PSI platform of 1968 contained two items that targeted families. The first related to women's rights in paid work and was coupled with the issue of the care of children. They espoused the cause of public child care services and social services for the family, which 'make it possible for women to be integrated in paid work and in society' (PSI, 1968: 6). The second family issue in the PSI's platform was related to legal changes. They supported the introduction of divorce and advocated the abolition of fathers' strong authority over the freedom of their children (PSI, 1968).

Throughout the 1970s the DC and the PSI continued to support the aforementioned requests, while 'lay and liberal' parties began to ask for explicit changes in family law. The latter had expressed their firm opposition to the Catholic Church's attempts to interfere in the nation's political life since the forming of the Constitution.

The Communist Party avoided addressing laws and policies concerning family issues. When it did so, it conceptualized the issues in terms of women's issues, women's rights, and/or individual freedom. In fact, it was only in the middle of the 1970s that the PCI directed its attention to the public refunding of day-care services, to the rights of working women and to women's choices about motherhood (PCI, 1976, 1979). The PCI began to include specific women's (family) issues in its platform only after the introduction of the day-care law, the maternity law, and divorce and family law reforms. While they may not have always initiated these laws, they provided crucial support. In addition, in the context of feminist movement activities, they sided with the movement, declaring support for the feminist struggles for women's liberation and freedom in areas such as divorce and abortion. It is likely that the controversial nature of issues related to social reproduction and caring work, set in the context of a highly ideologically polarized political party system, explains the gap between the party's manifesto and its support for these concerns. The party probably could not afford to take a position on such highly controversial issues.

The rediscovery of the family during the welfare state crisis In the mid-1980s political parties greatly changed their analysis of the position of the family. Since then, almost all political parties have framed issues explicitly in terms of family policy. In addition, broad debates on family changes and family policies have been raised at different institutional levels.

One of the main consequences of the welfare state crisis has been the rediscovery of the importance of the family in providing social protection and caring services. In the search for a solution to the welfare state crisis, the emphasis has been put on finding the best 'welfare mix' (a mix of public and private resources) and on how to better redefine social solidarity (Paci, 1989; Ferrera, 1993). The debate on the welfare mix has increased expectations that families (that is, women) should take most of the workload for the care of children and the elderly.

It is exactly in light of the welfare state crisis that new interest in the family in the Italian political arena can be explained. In the period 1987–90 a considerable number of bills were presented in parliament (Colozzi and Matteini, 1991). Nearly all political parties, several of the regions, and the government presented bills. Although most of them addressed the reorganization of social services and the definition of a national framework for social assistance, all explicitly pointed to the family as an actor in social policy.

The DC platforms were the first to explicitly request placing the family at the centre of social policy (*la famiglia al centro della politica sociale*) (DC, 1987). In 1992 they requested the approval of a 'family policy' (*una politica per la famiglia*) (DC, 1992). However, the position of the DC on state–family relations, and on women's primary role as mothers, has remained almost unchanged. Furthermore, families, in light of the welfare state crisis, were regarded as 'active' players in social policy and fundamental contributors to the optimal welfare mix. According to the DC's political party platform, the state should confine its policy interventions to maintaining and supporting family formation and stability, and/or to cases of emergency.

The 1992 political party platform of the PSI also addressed social policies for families. The PCI/PDS party was late in addressing social policies for families. More precisely, the PDS waited until the 1994 electoral programme. At that point it named families as the best producers of caring and service work for individuals (PDS, 1992). The long absence of the family from the PCI/PDS's political platforms does not mean that this party was not interested in issues relating to family relations or family responsibilities. In fact, the Italian regions and cities with a long 'red' cultural tradition tend to have higher coverage and better day-care services. In addition, the PDS proposed a bill entitled 'Women's Time and City Times' in 1989. The bill intended to create equal opportunities in the Italian welfare state based on family responsibilities and to make the use of time flexible throughout a woman's working life. The law was not passed. Nevertheless, the debate on 'City Times' and the reconciliation of urban times have become important issues in women's political debates and in debates about combining family responsibilities and work time. Since then, several regions have approved laws on 'City Time', and many Italian cities have introduced 'City Times' programmes (Bimbi, 1999).

The specific issues raised by the new debates on family changes and family policies have also had some response at the regional level. Between 1989 and 1995 six regions passed laws on family policy and others have proposals pending (Bimbi, 1999). The possible direction of family policies has also raised interests and debates in other institutional contexts (Comitato per l'anno internazionale della Famiglia, 1994). At the national level, the Ministry of 'Social Solidarity' was established in the 1990s. This is responsible for family policy, including a wide array of issues: minors, the elderly, the disabled, the voluntary sector and immigrants.

The Division between Left and Right in Spanish Politics

After the downfall of the dictatorship, Spain avoided developing both an explicit family policy system and social policies for family responsibilities. An explanation for this has been found in the legacy of the dictatorial regime. In the post-Franco era any form of social support for families has been identified, even more than in the Italian case, with the pro-natalist and conservative dictatorial family policy system, and has therefore been refused (Meil, 1994; Valiente, 1996a; Alberdi, 1997).

The Spanish feminist movement contributed greatly to this refusal. At the beginning of the 1970s the main claims and struggles of the feminist movement were for individual freedom and women's equal rights. These requests were coupled with a strong opposition to the pro-natalist and anti-feminist policies of the authoritarian regime. The movement attacked the institution of the 'family' more strongly than in other countries because it was regarded as the main source of women's oppression (Escario *et al.*, 1996).

Spanish society's cultural and political break with the past regime was reflected in a very broad legislative process, which deeply influenced family relations. One of the first and most symptomatic changes was the decriminalization of adultery and cohabitation, which took place in 1978.[6] Legislation concerning the use and advertisement of contraceptives occurred in the same year. Sterilization was deemed an offence until 1983 (Picontó-Novales, 1996).

The constitutional imperatives left their imprint on the family law reforms of 1981. The 1981 civil code reform[7] and the 1981 divorce law[8] changed the view of legal regulation in the sphere of family relations as well as assumptions about the intervention of the state in this field. The majority of feminist activists during the democratic period have focused their political activity in the area of legal changes in family relations. They have presented their interests as issues of *women's* policy, as opposed to *family* policy (Valiente, 1996a; Flaquer, 1998). Thus, legal changes were achieved in Spain under the headings of individual freedom, women's policy (i.e. equality between men and women) and/or in the name of child welfare, but never under the heading of family policy.

The Francoist ideology of the family and the feminists' requests for women's policy have jointly contributed to the building of a sort of inhibition about political or public debates on 'the family' and towards any social policies for family responsibilities.

Another reason for the delay in the development of family policies in Spain could be the absence of a national Christian Democrat Party, as well as a lack of family organizations, neither of which emerged during the transition to democracy (Giner and Sarasa, 1992). Finally, and more importantly, the Catholic Church did not advance demands for a 'family policy'.

The Spanish Constitution mentions the family as an institution to be protected, but does not explain which model of the family should be protected. During the rise of democracy in the 1990s, family benefits were not re-evaluated and social policies related to the family consisted of fragmented and 'poor' interventions. This approach has left responsibility for the reproductive and caring work completely to the family and to the kin group.

Throughout the period under analysis, the most important differences in how family issues are addressed have been between the centre-right and the left-wing parties. The analysis of political party programmes reveals that the lack of social policies for family responsibilities was also due to the coexistence of different ideological representations of the family.

The supporter of 'the family unit' Since the transition to democracy, centre-right Spanish parties such as the Unión de Centro Democrático (UCD) (1977, 1979) and the Coalición Democrática (CD) (1979) have strongly emphasized the institutional value of the family in their political programmes. More importantly, since 1979 centre-right party platforms have framed issues explicitly in terms of the family; that is, they have had no problems framing legal changes and policy interventions in terms of family issues. According to these parties' ideologies, not only should the state conceive of the family as the basic institution of society, but it should also help families to accomplish their duties.

At the end of the 1970s and the beginning of the 1980s, Spain experienced both a delicate political transition and a sharp economic crisis. Given that the UCD had the mandate from 1977 to 1982, one would expect to find great demand for social policies for family responsibilities in the political agenda. Instead, the funding for family allowances continued to be cut back substantially in real terms during this period. This was not due to any policy changes, but simply to the fact that the UCD government did not update the level of family benefits. According to Meil (1994), the entire period between 1976 and 1982 was marked by a lack of parliamentary debates and by an absence of political party concerns over the issue of protecting families with children.

During the 1980s the main Spanish centre-right parties, the Centro Democrático y Social (CDS) and Alianza Popular/Partido Popular (AP/PP), upheld in their party platforms the ideological representation of 'the family' as the main institution in society. They continued to address family issues in terms of both legal intervention and policy intervention. All through the period, these two political parties devoted whole sections of their programmes to the family.

The AP/PP – the main party in opposition to PSOE from 1982 – explicitly supported the introduction of a family policy system. In 1982, Alianza Popular requested the creation of a Ministry of the Family, Youth and Sport (Ministerio de la Familia, Juventud y Deporte). This ministry was intended to co-ordinate the public interventions devoted to protecting family unity, family functioning and family members. According to the AP/PP, 'family policy' must be composed of a wide set of state interventions that generally encompass the legal, economic and social supports that contribute to family unity. They requested detailed social policy interventions that supported motherhood, marriage, family unity, family care for disabled members, and women's employment (AP, 1982).

In its 1986 political programmes, the PP reaffirmed that the family (the one based on marriage) must be protected and maintained (*conservada*) (PP, 1986). The PP also expressed its ideas about the division of responsibilities between the state and the family. They suggested that the state must intervene only if the family has failed in its functions. Furthermore, they argued that the family ought to be protected legally (i.e. life rights), socially (i.e. housing rights, educational choice, survivors' pensions), and financially (i.e. agricultural families' enterprises, *empresas familiares*).

The supporter of individual rights and of 'Women's Policy' In left-oriented party platforms, declarations and discourses on the family, as well as specific issues framed in terms of the family, have been altogether avoided. Rather, left-wing parties tended to stress the individual, his or her social rights, women's policy, and the redistribution of wealth among social classes.

In the 1980s Spain experienced a period of democratic consolidation. New issues concerning personal and individual rights (abortion, equality between the sexes, divorce) and new demands for social rights (universal pension and health rights) were raised in society and strongly supported by left-oriented parties.

In general, the analysis of political party platforms during this period reveals that the PSOE and the PCE conceptualized family relations and family responsibilities in terms of gender issues. Thus, demands for legal changes were framed in terms of gender equality, and demands for social policies in terms of *women's* policy. These two parties also gave high priority to the issue of women's emancipation. This emphasis on women's emancipation and women's rights was similar to that held by the Italian left-oriented parties.

Throughout the 1980s the PSOE kept away from rhetoric concerning the institutional dimension of the family and avoided framing family issues. Despite this, it included social policies in its platform that targeted assistance at certain categories of 'needy' families (low-income families, families with disabled children). These interventions were suggested within the framework of social insurance reform (PSOE, 1982).

In the mid-1980s the socialist parties raised the issue of family responsibilities in their platforms. However, the requests for financially assisting low-income families were conceived of as measures to achieve equity between social classes. In the PSOE electoral platform of 1989 the party devotes a subsection to family benefits (*prestaciones familiares*) for low-income families. The preferences of socialists for specific interventions to targeted families (families in need) were translated in 1990 into the current Spanish system of means-tested benefits for families.

The issue of combining work and family appeared in Socialist Party platforms only in 1989. This issue was related to care services for children. While these services were initially seen in terms of *women's* policy, they were later conceptualized as educational policies aimed at reducing social inequalities among infants and children up to the age of 6 (PSOE, 1989). The request for a new educational policy was translated into law in 1990.

In sum, the PSOE (and the PCE) focused until the end of the 1980s on women's inequalities in society, on legal and social changes for achieving gender equality, and on social security benefits for low-income groups. They never directly addressed issues of the family. They did not make any reference to the individual as potentially bound to the family unit. This existed only in terms of fiscal policies.

During this period of PSOE governance, the division between centre-right and left-oriented parties in their attention towards the family and towards family issues remained substantially the same. The left-wing political parties did not mention the family because it was regarded as a conservative institution. The past regime's ideology of the family and the feminist movement's struggle for women's policy contributed to the diminishing importance of the family as an institution for the most progressive political forces in Spain. The left-oriented political parties accepted without any criticism the feminist elaboration of 'the family', which they depicted as the main institution oppressing women.

The lack of discourse on family issues in the Socialist Party platform has to be seen in relation to the position supported by the feminist movements. The main feminist institution of the central state which existed in 1983, the Institute of Women (Instituto de la Mujer, IM), rejected Francoist family policy and avoided framing child care issues as family issues for a long time. Its concern was mainly *women's* policy and gender equality.

Importantly, the IM did not maintain a department dedicated to family affairs. In the first Equality Plan programmes (IM, 1987), social policy reforms were mentioned, but with the goal of equality between men and women. No mention was made of the division of responsibility between the state and the family in the areas of caring work or financial support for families with children. In other words, during the1980s the IM in Spain ignored the issue of child care for some of the same reasons feminists did so. In general, however, it has to be said that the IM and feminists were overwhelmed by the demands of the other issues they decided to support (Valiente, 1995).

In the IM's second Equality Plan (1993), proposed five years later, a section was devoted to the theme of the equal division of labour in the family between men and women in the domestic sphere. At last, the issue of the care of children was framed, but with the goal of understanding the socialization process of male and female children within the household. It was also in 1993 that the Spanish Institute of Women advanced their first demands for increasing public child care services (IM, 1993, section 5.1.1).

During the 1990s family issues drew new attention and began to find a place in the Spanish political agenda. For instance, in 1994, the United Nation's International Year of the Family, the Spanish parliament created a commission to study family changes and the forms of state interventions. At this time, one subsection of the Ministry for Social Affairs (Ministerio de Asuntos Sociales) changed its name from the General Direction for the Protection of Minors (Dirección General de Protección Jurídica del Menor) to the General Direction for Minors and the Family (Dirección General del Menor y de la Familia). This change was not only symbolic: it was Spain's first attempt to overcome the inconsistencies of a public administration which dealt with children, youth, the elderly and women without ever relating them to each other in terms of families.

Legal Definitions of Family Relations

As I have shown, both Italy and Spain proclaimed the formal principle of sexual equality in their constitutions, and this was also accomplished in several other countries (i.e. Germany in 1949, and France in 1946 and 1958). Although the Italian Constitution itself contains the formal principle of sexual equality, it appears ambiguous in some other parts (i.e. concerning the representation of woman's role as first mother and wife and then as worker). The tension between the different principles allowed judges to interpret them differently according to gender, and, in cases related to intergenerational issues, according to the status of children born within or out of wedlock. Until the 1970s the constitutional principles continued to exist side by side with a civil code that institutionalized the inequality of married women (Del Re, 1995). Therefore, although Italy changed its institutional–political framework after fascism and the Second World War, the traditional values and norms defined by the Catholic Church and codified during fascism (i.e., *Codice Rocco* was still the only code ruling family relations) continued to strongly influence the family and family relations. The configuration of a strong Christian Democrat Party, the main governmental force since the 1948 election, and an active Catholic Church, which had always interfered in family matters, favoured the permanence of traditional family patterns, setting back family law reform by almost 30 years.

However, in most European countries the edifice of traditional family law (based on the principle of legitimacy and on marital and paternal authority) remained standing until the 1960s. Although Spain declared the formal principle of sexual equality much later (1978) than did Italy, it soon followed other European countries' reforms in family law and approved a new civil code (1981) which profoundly changed the legal definitions of family relationships. Thus, despite the 30-year gap between Italy and Spain's political democratization, the democratization of family relations as well as the introduction of several measures representing individual and personal freedom (i.e. divorce, abortion, contraception, etc.) occurred almost during the same period. In the following sections I explore family law, that is, regulation, rights and obligations of family members to each other, as they are found in the civil code and in some legal norms. I focus, for instance, on divorce and adoption laws. The limits of the analysis should be spelled out: it does not look at all at the implementation of the law; nor does it provide an illustration of what happens when the family obligations prescribed by law are not enforced.

Definitions of Gender, Intergenerational and Kin Relations in Italian Law

Gender relations and obligations There were two main family law reforms during the new Italian republic. The first was the introduction of divorce in 1970, which signalled a revolution of the traditional concept of marriage, by which 'only the death of one spouse could break down the marriage', towards a principle inspired by the idea of a revocable contract of marriage. The Italian divorce law, late compared to those of other western countries, benefited from the legislative developments that had already been undertaken elsewhere. As a matter of fact, the Italian divorce law fully supports the idea of 'divorce remedy', in which divorce can be 'no fault' and not perceived, as suggested in the *Code Napoléon*, as a punishment (Vincenzi Amato, 1988; Barbagli, 1990). According to the law, the claim that living together had become intolerable is sufficient grounds for divorce. Despite this, Italian law discourages the dissolution of marriage, and divorce is simply the final legal step in a longer process of separation. Divorce petitions in Italy can be accepted only after the spouses have been legally separated. This mandatory separation period was originally set at five years, but was reduced to three years in 1987.

Finally, in 1975 family law reform took place. This reform marked the second important revolution with respect to gender and intergenerational relations. The reform law abolished the anachronistic legal concept of 'family head', giving both spouses equal authority. Couples who enter into the marriage contract have the same rights and the same duties. Specifically, parents have mutual obligations. In addition, they must contribute to the needs of the family according to their capacity and their ability to work both in and outside the home.

As far as the division of property is concerned, married couples typically hold property in common (*comunione dei beni*), with explicit acknowledgement of the value of unpaid family work. Couples do have the option of keeping their property separate if they prefer (Saraceno, 1998).

Intergenerational obligations The Italian child legal system, first with the 1967 adoption law and then with the 1975 family law reform, is inspired entirely by the principle of 'the child's best interests', '*primauté de l'intérêt de l'enfant*', '*nell'interesse del minore*'. With this principle the legislature intended first of all to ensure children's welfare through their right to be placed in a proper and favourable family environment. Through the 1975 family law reform, the holder of parental authority (*patria potestà*) is no longer the 'family head', but both spouses equally and together.

Parents throughout Europe are bound by law to support their children. In continental Europe grandparents and sometimes, as in the Italian case (see Articles 433 and 439), brothers and sisters can be obligated to provide child support (Dopffel, 1988). According to Article 147 (*doveri verso i figli*) of the Italian civil code, parents have obligations concerning the maintenance, (*mantenimento*), upbringing, and education of their children. However, unlike other European states, the Italian state does not intervene immediately if parents have insufficient resources to meet these obligations. First, it obliges their ascendants (*ascendenti*) to provide the parents with the necessary means for the children (cf. Article 148 of the civil code) (Pocar and Ronfani, 1995). Finally, according to Article 439 of the Italian civil code, included under the heading '*alimenti*', siblings' legal obligations are limited to the primary needs, except in the case of a minor in need. In this case legal duties extend to educational support.

Kin obligations The Italian family law reform of 1975 maintained legal financial obligations among kin, although it abolished, for instance, the support duty of grandchildren towards their grandparents. This is a norm dating back to 1865, which was subsequently reformed in the fascist civil code of 1942. According to the current Italian civil code (*Degli alimenti, titolo XIII*) kin members are obliged to provide financial support for each other in proportion to their income, rather than degree of kinship. However, kin support duty is given only in the case of need. The financial support of kin is given depending upon the necessity of the needy person and the economic capacity of the givers. According to the civil code (Article 433) the kin 'with responsibility for maintenance' (*parenti tenuti agli alimenti*) obliged to provide income support are: spouses; legitimate and illegitimate children; parents, or in their absence, next of kin '*ascendenti*'; sons and daughters in-law; parents in-law; siblings and half-siblings. Siblings bear more obligation than half-siblings, and Article 439 of the civil code states that between siblings the support duty cannot go beyond the strict necessity.

As can be easily observed, the kin obligation to provide financial support for each other goes well beyond the nuclear family and the residential boundary of the household. The assumptions contained in kin legal obligations are two-fold. First, the legal acknowledgement of the importance of family solidarity, or better, of family/kinship solidarity (Vincenzi Amato, 1973). Second, the survival of kin obligations points to the role still played by the principle of 'subsidiarity', that is, the role of the state is regarded as subsidiary to that of the family.

In practice, kin obligations become fundamental when an individual requests social assistance or if a frail elderly person has to be hosted in a publicly run institution or funded house. In both cases the state will not assume the financial burden for the family/kinship dependent members until the means of all these kin, who are legally responsible for maintenance, have been thoroughly investigated (Auletta, 1984; Longo, 1985; Trifiletti, 1995a). However, the issue of the relationship between social assistance and family obligations has not been clearly defined either in the Italian Constitution (Article 38) or in the subsequent social assistance and social security laws (i.e. social pension) (cf. Longo, 1985). In Italy, a lack of a nationally defined minimum income and a minimal social pension scheme have contributed to a lack of clarification of what the relationship between family solidarity (i.e. family and kin legal obligations) and social solidarity (i.e. public provisions) should be.

The same principles and ideas of solidarity and subsidiarity of the state in relation to the family are found in Article 12 of the adoption law (*legge sulla adozione*), which obliges kin up to the fourth degree to take care of the adopted child (Auletta, 1984).

Definitions of Gender, Intergenerational and Kin Relations in Spanish Law

The 1981 Spanish civil code reinforced the principle of equality between spouses, breaking completely with the past tradition of subordination of the wife to the husband. Husbands and wives are viewed as equals in rights and duties. Joint parental authority replaced pre-constitutional paternal authority. The principle of financial equality and solidarity was also respected, obliging both spouses to contribute to the support of the household (Alberdi, 1992). According to the civil code, the husband and wife share management of communal properties, unless they select a different marital property regime. This differs from the situation prior to 1981, when only the husband managed such property. In addition, according to Article 1438 of the civil code, the unpaid housework and caring work can be acknowledged as a contribution to the support of the household, and care-givers might have rights to compensation in the case of separation or divorce (Alberdi, 1994).

As far as the form of marriage is concerned, the two main peculiarities emphasized in the civil code are the secularization of marriage and the introduction of divorce. Civil marriage disappeared as a subsidiary form and the state introduced the dual system of 'civil' matrimony and 'canonical' matrimony with civil effect. This system was established on the principle of choice, whereby a Spaniard can choose the marriage law closest to his/her attitude and preferences without having to declare his/her own religion.

According to the guidelines given by the Constitution, the marriage bond can be dissolved according to the desire of one or both of the spouses. Following the constitutional principles and after a lively parliamentary debate, the divorce law was finally approved in 1981. According to Spanish divorce law, a separation must precede the granting of divorce. A couple must be separated for a minimum period (generally one year) and only after separation can divorce be obtained. Mutual agreement is not necessary for divorce, but in fact if both partners agree, divorce is easier and more rapid (Alberdi, 1992).

Intergenerational relations and obligations As mentioned earlier, the Spanish Constitution (Article 39, 2) states that all children are equal before the law, thus eliminating discrimination according to a child's birth status. The civil code reinforces such equality by establishing that all children have the same rights within the family, regardless of whether they are biological or adopted, and whether they are born within or outside of marriage (Article 108 of the civil code). The civil code admits a broad possibility for investigating paternity and maternity.

Parents' obligations are stated in Article 154, 'to take care of them, keeping them in company, to support them, to provide them with education and with integral formation' (Alberdi, 1992).

Relations between kin The obligation to provide financial support for kin (*la obligación de alimentos*) is also included, as in the Italian case, in the Spanish civil code. This seems to be a peculiar legal institution of three Mediterranean countries: Italy, Spain and Portugal (Millar and Warman, 1996b). Apart from the abolition of the legal differences between legitimate and illegitimate children, Spain's reformed civil code (1981) maintains almost untouched the previous legal obligations between kin.

According to Spanish civil code (Article 142 and following), when social policies and public measures insufficiently provide a dependent adult with the necessary commodities, certain members of the family must financially support these dependent adults.

According to Article 144 of the Spanish civil code, spouses, ascendants (*ascendientes*), descendants (*descendientes*) and siblings (to a limited extent) are required to meet these mutual obligations of support. Spanish law differs from Italian civil law in that it reserves certain levels of support for certain levels of kinship. Spouses, *ascendientes* and *descendientes* must reciprocally provide 'broad support' (*alimentos amplios*), or all the provisions necessary to maintain one's particular standard of living. Brothers and sisters are only obliged to provide 'restricted support' (*alimentos restringidos*), or provision for basic needs (*auxilios necesarios para la vida*). Both types of support are obligatory only in cases of necessity (Diez-Pícazo and Gullón, 1990). It has to be observed that Spain, unlike Italy, does not mandate kin obligations between a person and his or her in-laws.

This civil law best demonstrates the institutionalization and legal obligation of family/kinship solidarity in Spain. Indeed, the duty towards kin is not just a civil obligation; failure to comply with these duties may lead to criminal prosecution. While Spain has these legal obligations, the state assists kin providers with their legal obligations by compensating them through tax relief. For instance, a tax deduction would be granted to a tax payer who provides for an elderly, low-income ascendant.

NOTES

1. The Moncloa Pacts were a series of economic and social agreements reached by the main leaders of the Spanish political parties, aimed at reinforcing the recent democracy and overcoming the economic and social crisis of that period.
2. In 1973 the leader of the PCI, Enrico Berlinguer, launched the 'historic compromise' between the three main political parties, the PCI, DC and PSI. The purpose of this, according to Berlinguer's plan, was to find a meeting point between the Catholic and communist ideologies (Ginsborg, 1989).
3. The UCD was a centrist party with no clear ideological definition. The UCD's lack of both organizational consolidation and a well-defined and differentiated ideological party programme resulted in its dramatic collapse in the 1982 election. Afterwards, a part of the UCD joined the PSOE and a part joined the AP. The Christian Democrats formed their own party, the Partido Democratico Popular (PDP), without influence, which later formed a coalition with the AP. In 1982 Adolfo Suarez set up the CDS, which afterwards entered into the 'constitutional bloc' with the PSOE. The Alianza Popular (AP) on the right first reconstituted itself as the Coalición Democratica, and then as the Coalición Popular (between 1979 and 1986). Since 1988 it has called itself the Partido Popular (PP).
4. For the Italian case the analysis of political party platforms takes place from 1948 to 1992. The complete collection of data was provided by the WZB (Wissenschaftszentrum Berlin fur Sozialforschung gGmbH) of Berlin. This data collection includes original political party platforms for all Italian political parties since 1948. Some political platforms are missing because the party refused to give them or because they did not exist for certain elections (especially during the electoral campaigns of the 1950s and the 1960s). I use newspaper and journal articles to replace the missing political party platforms. For the Spanish case, the analysis of political party platforms goes from 1977 to 1993; data includes all original national political parties' programmes.

5. Law no. 151, 19 May 1975, 'Family Law Reform'.
6. Law no. 2, 26 May 1978.
7. Law no. 11, 13 May 1981.
8. Law no. 30, 7 July 1981.

Chapter 6

FAMILY CHANGES AND FAMILY SOLIDARITY

This chapter aims to provide a general picture of family changes and family solidarity in Italy and Spain. In particular, it focuses on how welfare state institutions, socio-economic structures and family contexts have changed and how they mutually interact. The analysis covers the period from the 1960s to the 1990s, that is, the period for which more comparable and reliable data are available. Bearing in mind the differences which exist in the institutional and political settings of the two countries, I will try to look at their developments within the European context. I can thus illustrate how Italy and Spain, with their unique patterns of family change, modernization and welfare state development, support the idea of a 'southern model'.

The analysis illustrates how Italian and Spanish families present indicators of both deep change and continuity. The increase in women's employment in the 'official' labour market, the rapid decline in fertility and the increase in conjugal instability are surely indicators of changes. However, marriage as an institution is still very important and, as a result, birth outside marriage is still rare and divorce is not yet so widespread as in other European countries. Cohabitation and living as a single young person are still very unusual phenomena. Furthermore, young people tend to delay leaving the parental home. Finally, and most importantly, in these societies intergenerational family and kin ties are still very important. This solidarity and the importance of welfare functions provided by family and kin members cannot be explained by the nuclearization of the family. Rather, this model points to the hegemony of a shared cultural family model in which intergenerational and family/kinship solidarity are expected and are fundamental in providing emotional, economic and social support.

Modernization Processes and Welfare State Expansion

In the post-war period, the occupational structures of Italy and Spain were characterized by the prevalence of the agricultural sector until the end of the 1950s in Italy (see Figure 1) and until the mid-1960s in Spain (see Figure

3). Moreover, in Spain the decline in the number of agricultural workers was achieved without prior peaks of industrial workers – there was a jump from an agrarian to a service-oriented economy. In both countries the process of modernization, as measured by the decline of agricultural production as a percentage of the gross domestic product (GDP), the level of GDP per head, and the level of total social expenditure was very rapid. Modernization in both of these countries was accompanied by structural employment shortages and deep territorial imbalances, which resulted in great internal and external migrations from the 1960s until the mid-1970s. This period of change was also characterized by high rates of self-employment and a higher rate of workers employed in small enterprises within agricultural settings. Finally, there was an increase in the number of people (mainly women) working in the informal or 'black sector'.

With the exception of maternity reform in 1950 and the pension reform of 1952, the Italian welfare state did not change much during the first two decades of the new republic. Consequently, the Italian system maintained the institutional principles and regulations, as well as the organizational and administrative patterns, of the past regime. Consequently, the welfare state remained a fragmented system characterized by marginal adjustments and clientelistic exchange (Ferrera, 1984).

The Italian welfare state grew most during the 1970s and early 1980s, finally reaching European levels of development. In 1981 total public expenditure as a percentage of the GDP passed the threshold of 50 per cent, while the total social expenditure reached 29.1 per cent of GDP. As Ferrera (1984) shows, Italy has demonstrated average levels of total public expenditure and growth since the 1960s. The real process of welfare state reforms and secularization of the society began during the 1970s. Administrative decentralization gave regional and local authorities the competence for health care (in 1978 the National Health Service was established), social assistance, housing and vocational training policies. This administrative decentralization occurred during the period of maximum expansion of social and caring services (kindergartens, day-care for children under the age of 3, home help for the elderly, etc.).

In Spain, the social security system provided employment-related contributory benefits for retirement, disability, sickness and widowhood by the end of Franco's dictatorship (1975). The insurance scheme covered some 75 per cent of the population. Different groups of workers received different benefits, and there were major divisions between the rights available to different groups in the occupational hierarchy (Almeda and Sarasa, 1996). The democratic regime inherited these policies and continued the expansion already begun by the end of the 1960s. Subsequent democratic governments have expanded social protection measures,

Figure 5
INDICATORS OF MODERNIZATION AND WELFARE STATE EXPANSION, SPAIN AND ITALY, 1961–91

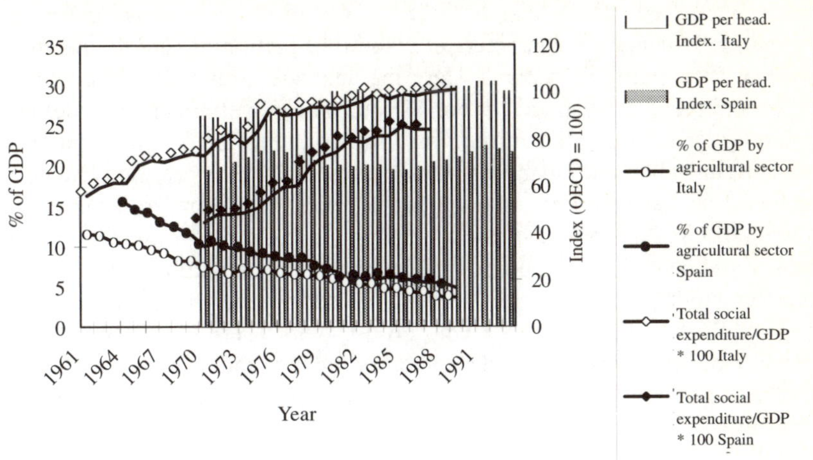

Sources: For percentage of GDP per sector: OECD (1980, 1989 and 1991). MZES (1994). IC 1. For total social expenditure: Alcaide Inchausti (1988), tables 24, 25 and Ferrera (1984), table 4. For GDP per head: OECD (1995), p. 148.

extended pension coverage, and introduced universal unemployment benefits (1990). The National Health Institute (INSALUD) was set up in 1982, creating a 'near' universal health system. Welfare state expansion during the 1970s and the 1980s resulted in a huge growth in social expenditure which passed from 36.2 per cent of total public expenditure in 1960 to 62.1 per cent in 1975, to 67.1 per cent in 1985 (Rodríguez Cabrero, 1989, table 3). However, although social expenditure as a percentage of GDP increased very quickly at the end of the 1970s, it was still under 20 per cent in the 1980s and thus still below the European average level (cf. Eurostat, 1993).

Like Italy, Spain also experienced changes in the institutional configuration of the welfare state with the transition to democracy. There was a gradual administrative decentralization from the central government to the Autonomous Communities and local administrations, especially in the field of health services, educational responsibilities, social services and means-tested 'minimum family income' benefits (Almeda and Sarasa, 1996; Moreno and Arriba, 1998).

The Current 'Dualistic' Protection System

Despite the different timing of welfare state expansion in the two countries, there are several common characteristics relating to how the two welfare states functioned. These can be seen as the result of similar interplays between labour market characteristics, family functioning and the peculiar mode of their political systems. Here the analysis turns to the impact that certain unique characteristics of the Mediterranean welfare state model has on family changes and the current institutional definition of family models. The current Italian and Spanish welfare states, as in the 'corporatist model', are based on 'occupational status' (that is, social rights are granted mainly to workers) and are oriented towards income transfers. But these countries have implemented a 'dualistic' system of protection: high protection for whoever works in a stable job, that is, the adult male (breadwinner), and very weak protection for those who do not work (Ferrera, 1996; Guillén, 1996). How has this two-fold protection system interacted with family changes and how has it contributed to the permanence of family kin relations?

Female Participation in the Labour Market

In order to understand the organization of family life, it is important to have an idea about women's employment patterns and how female labour force participation is encouraged or discouraged by the welfare state. In the post-war period, Italy and Spain had very low rates of female labour force participation, with Spanish women least likely to be employed. During the 1950s and until the mid-1960s female activity rates increased in Spain (cf. Figure 4), while in Italy during the 1950s and 1960s, as a consequence of the first phase of industrialization, more and more women returned to traditional roles as housewives at the time of their marriage or upon the birth of their first child.

Since the beginning of the 1970s, female labour force participation has increased in both countries, with greater velocity in Italy than in Spain (see Figure 6). What has been really new for both countries, however, is that the female labour supply is more related to women's higher education levels and desire to work, than it is to market demand.

However, as Figure 6 clearly shows, despite the rapid growth of female activity rates, Italy and Spain have still not been able to close the gap with other European countries. Structural employment shortages are one important historical feature of both countries that surely has shaped the contrast between southern and central/northern European women's labour force participation. In addition, regional differences in the employment

Figure 6
FEMALE ACTIVITY RATE, 1960–94

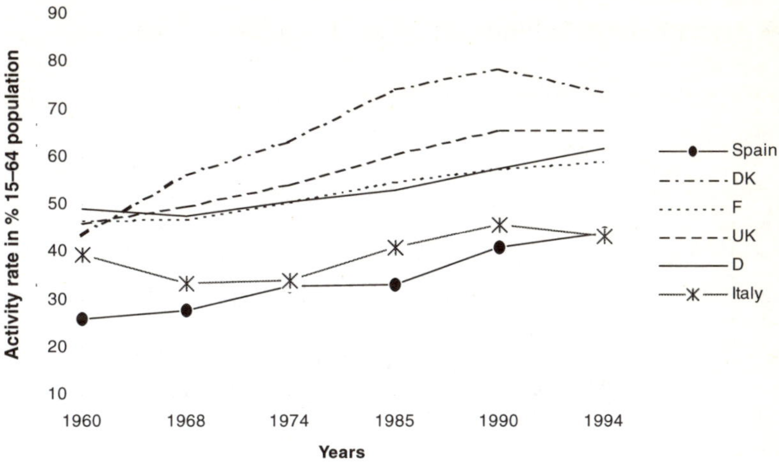

Source: OECD (1996), *Historical Statistics*, 1960–1994, table 2.8 (data refer to female labour force as % of female population 15 to 64 years).

structure were and still are important. Despite the rapid augmentation of southern female activity rates from around 34 per cent in the mid-1970s to around 40–45 per cent in the 1990s, they are still the lowest among the analysed European countries. Aside from this difference, women in Spanish and Italian labour markets are integrated in different ways than central/northern European markets.

One of the main characteristics of Spanish and Italian labour markets is the high presence of 'black' or 'grey' jobs. These jobs, mainly held by women, typically offer minimal social security, unpredictable or irregular hours, and relatively lower pay. Informal sector employment, which can include seasonal agricultural labour and domestic work, is not included in official labour statistics and women's labour force participation is consequently often underestimated (Muro *et al.*, 1988; Solsona, 1991; Balbo, 1976; Paci, 1980).

There are three other peculiarities marking the recent entrance of women into the labour market in Italy and Spain.

1. The augmentation of female activity rates meant and still means an increase in women's unemployment rates. The gender division of

Figure 7
FEMALE EMPLOYMENT AND UNEMPLOYMENT, 1989

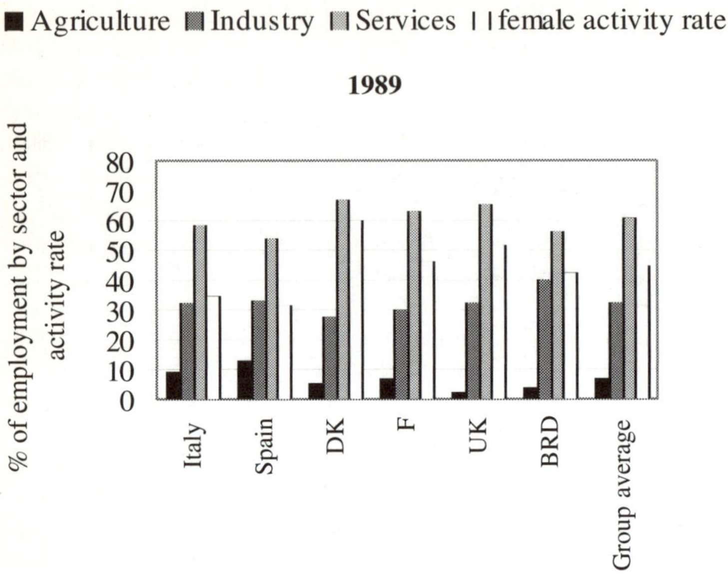

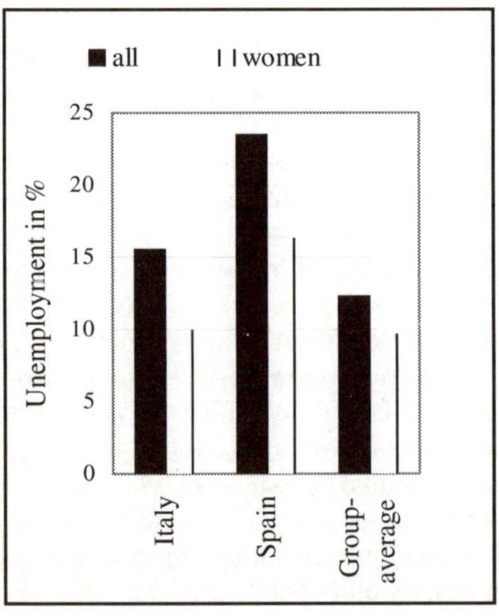

Source: Eurostat (1992a), p. 41; (1991), p. 96. The international comparison of unemployment is difficult; therefore this survey is restricted to comparable data from the Community Labour Force Survey.

unemployment is much more clear-cut in Italy and Spain than in other countries (cf. Figure 7).
2. Women's part-time employment is very low. In 1989 10.9 per cent of Italian women in the formal labour market were employed part-time versus 11.9 per cent in Spain. In France the part-time rate was 23.8 per cent; in West Germany, 30.7 per cent; in Denmark, 40.1 per cent; and in the United Kingdom 43.6 per cent (Eurostat, 1991; Eurobarometer, 1991a: 34).
3. Italy and Spain, as shown in Figure 7, have less-developed service sectors, which are important sources of employment for women in other countries. In contrast, Italy and Spain have a relatively higher proportion of employed people in the agricultural sector.

Southern European women also appear to have unique career patterns compared to other parts of Europe. A longitudinal study on employment and the family in the EU carried out for Eurobarometer in 1991 clearly shows how different labour market characteristics affect women in different ways. Women in Spain and Italy enter the labour market at later ages and at lower rates, they have a very high continuity rate, and they seldom work part-time. After the birth of their first child, a majority of married women continue to work full-time in Italy, whereas in Spain they either stop employment completely or continue full-time. By contrast, in West Germany and Great Britain, women have a high initial entry rate, but low continuity, punctuated by long interruptions and high reliance on part-time work. A very high number of women leave the labour force after the birth of the first child (Eurobarometer, 1991a). Other national and cross-national studies have also shown the permanence of women in the labour market in Italy and Spain during their adult years (Ambrosini and Rossi, 1994).

However, as Bettio and Villa (2000) argue, what has really changed is not the life cycle pattern associated with motherhood, but the propensity to participate across cohorts. There is an increasing number of younger women who enter the labour force, while those in the central age group are still likely to leave employment. In addition, there are more women who show continuous participation over the life course. According to the authors, education has replaced motherhood as the most important predictor for participation or non-participation. In other words, female participation is much more influenced by the educational level of women than by the presence of children. A similar conclusion is suggested by another study (Bison, Pisati and Schizzerotto, 1996) in which the entrance and the permanence of Italian women in the labour market is mainly related to the level of education, civil status and the presence of children. These authors also note the importance of the geographical area of residence. For instance, highly educated women tend to have greater levels of work experience and a longer working career than do lower educated women. The chances of a better working career are reduced if the woman is married; even more so if she has a child. In addition, the probability of entering and remaining in the labour market increases if a woman is living in the centre-north of the country.

Family Changes and Family Continuity

According to many social analysts (Roussel, 1994; Barbagli, 1990; Barbagli and Saraceno, 1997) the great changes in western European family patterns began in 1965, as a consequence of both demographic changes and normative shifts. Since 1965 a reduction in fertility rates and an increase in divorce rates have taken place in most European countries, with varying intensity and timing. These family changes are related to the progressive diffusion of cohabitation and to the increase in births outside of marriage. During the 1970s these changes became more radical in Scandinavia and more widespread in central Europe, but remained weak in southern Europe. The biggest changes were in the relations between sexes and generations. Equal opportunities became an important issue in public life (Roussel, 1992), and changes in family roles also occurred. More women entered the labour market, in a wider range of occupations and for a longer period, than at any other time since the Industrial Revolution (Rapoport, 1989).

Despite the profound changes in family behaviour, the traditional forms of matrimonial and intergenerational ties, which point, in particular, to the importance of the family/kinship group, continue to be important in Italy and Spain. The sections that follow[1] examine changes and continuity in Italian and Spanish family behaviour during the three main phases of the family cycle using national level data.[2] First I will illustrate changes in the family formation phase (marriage patterns, cohabitation and being single), as well as the phenomenon of adult children remaining in the family home. Next I explore changes in the family enlargement phase (mean age of childbirth, number of children, births outside of wedlock). Finally, I point to the changes in family dissolution (forms and degree of marital breakdown). The concept of family life cycle is useful because family patterns vary not only geographically, but also in the same family unit over time.

The Family Formation Phase

The reduction in the number of marriages is one of the most important characteristics of recent Italian demographic developments (Golini, 1988). As can be seen in Figure 8, the great reduction in the number of marriages started in Italy in the 1970s, later than in other European countries. While it

Figure 8
GROSS MARRIAGE RATE, 1960–95

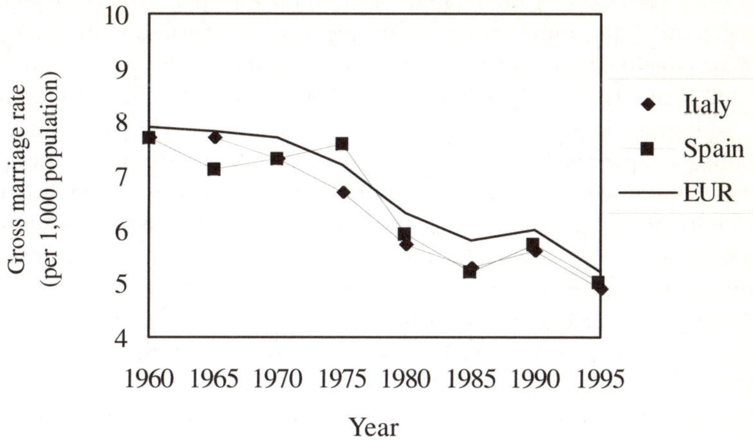

Source: Eurostat (1995), table F-3, p. 114.

is common to all regions, it is more typical of the centre and north (Golini, 1988). In Spain the gross marriage rate generally has decreased since 1981, stabilizing from 1986 to 1990. For 1982 one can also detect regional differences in Spain, but they do not seem to be very important or clearly structured (Delgado, 1993; Instituto de la Mujer, 1990; Eurostat, 1992b).

So far, gross marriage rates in Italy and Spain are below the European average. The data on the mean age of marriage for both Italy and Spain seem to support the hypothesis that the figures do not indicate a reduction in the number of marriages, but a delay in the timing. In fact, in Italy the decrease in the total number of marriages has especially affected young women. At the same time, the marriage rate of women over the age of 25 has increased (Golini and Menniti, 1994). There is also much variation at the territorial level, especially in female behaviour. The segment of the population that has changed most rapidly is women living in the big towns of the centre and in the north of the country (Golini, 1988). In Spain, too, the data show that since 1980 the percentage of people marrying before the age of 25 is decreasing. The age of women at the time of first marriage has risen since the 1980s in all European countries. During the 1990s it remained lower in Italy and Spain than in Denmark, Sweden and France, but higher than in Austria, Belgium and the United Kingdom (Eurostat, 1995, table F-15). In Spain and Italy a number of circumstances seem to favour the delay of marriage: high unemployment rates and increasingly precarious employment of young people, prolongation of education, expensive housing and cultural shifts after 1968 concerning sexual relations before marriage (Flaquer, 1994; Golini, 1988; Garrido Medina, 1993).

Another indicator of the change in the family formation phase (and of the degree of institutionalization of marriage) is the diffusion of non-marital living arrangements such as cohabitation and one-person households. In Italy and Spain existing data point to a low significance of the phenomenon of cohabitation (Ditch, Barnes *et al.*, 1996, table 1.9). The percentage of cohabiting couples in Italy is very low in comparison with other countries and has not shown significant variations in recent years.[3] Cohabitation is more common in Italy in the north-west, in urban areas and among higher-educated people and employed women. Moreover, Italy is unique in that most cohabiting couples are separated or divorced individuals rather than predominantly young people, as is the case in other countries. In Spain most cohabitants are young (18–34), non-practising Catholics, who live predominantly in urban areas in the Catalonia, Madrid and Basque country (Valero, 1992). The scarce prevalence of unmarried cohabitation in these two countries, and the low level of marital instability, are associated with a lower number of births outside of marriage.

Although the number of single-person households is increasing in Italy and Spain, very few people living alone are young. Rather, the elderly compose the clear majority living in this type of household. (Council of Europe, 1990; Eurobarometer, 1991b; Istat, 1993; INE, 1991).

The delay in life course phases Delayed marriages and comparatively low levels of cohabitation and single-person households among the youth mean that young people in Spain and Italy stay longer with their parents than in other countries. According to the comparative data provided by Eurostat (1997), in 1995, 87 per cent of young people aged 20 to 24 years lived with their parents in Italy, compared to 89 per cent in Spain, 55 per cent in West Germany and 47 per cent in the United Kingdom. The difference between Spain and Italy and the other European countries is even greater if we look at data for the 25–29 age group. Here we see that 56 per cent of young Italians (25–29) are still living with their parents, while 59 per cent of Spaniards do so. In contrast, only 21 per cent of young Germans are still living with their parents and even fewer young French and Danish remain in the family home.

The reason for young people's tendency to remain in their parents' home is not simply due to high levels of youth unemployment. Several Italian studies show that a substantial percentage of young people live in their parents' home even when they have a job (Menniti *et al.*, 1992; Sabbadini, 1999). According to Italian census data for 1991, 61.8 per cent of young adult males (18–29) with a job still lived with their parents. Such figures are even higher for the more prosperous regions of the centre/north of Italy than for the south, where higher unemployment rates are found (Piccone Stella, 1997). The lack of stable employment, affordable housing and more years spent in school can account only partially for this phenomenon.

The most important explanatory variable for late home-leaving in southern Europe is the cultural one. It is a cultural peculiarity of Italy and Spain that makes it acceptable and even desirable for young people to stay with their parents for a long time. Cavalli's surveys on this issue demonstrate that young people prefer to remain in their parents' home, where they receive economic support and social security along with personal freedom (Cavalli *et al.*, 1993). This phenomenon points to deep changes in traditional family authority. While the 1950s and 1960s were characterized by many conflicts and misunderstandings between parents and children, nowadays young people use their family of origin as a fortress within which to prepare for a good start in adulthood (Piccone Stella, 1997). In this new family context, young adults living in their parents' home enjoy higher levels of autonomy both in terms of sexual intimacy and individual freedom (Valero and Lence, 1995; Mingione, 1994). While Italians and Spaniards follow western European trends in terms of marrying later and doing so outside of the Church, they differ in that they do not start life away from the family home by cohabiting or living alone before they marry.

Family Enlargement Phase

Italy and Spain are considered late-comers in the demographic transition from high fertility to low fertility (Livi Bacci, 1977). Research undertaken over the last few decades shows that Spain and Italy demonstrate similar changes and features during the family expansion phase of family life. These changes resemble those of western Europe, but have been delayed by a few years. Italy's fertility decline began rapidly in the middle of the 1960s, preceding that of Spain. Spain rapidly caught up and reached similar fertility levels (see Figure 9). Similarities with Europe can be found in the increase of the mean age at first birth, in the reduction of childbearing among very young women, and in the stabilization of fertility rates after age 30.

There are two main characteristics of fertility patterns common to Italy and Spain but different from other western European countries. First, fertility patterns in Spain and Italy are quite heterogeneous across the regions. This differs from other western European countries, where fertility patterns are more homogeneous across different areas of a single country (Delgado and Livi-Bacci, 1992). Figure 9 illustrates the fertility rates of the two most disparate regions of Italy and Spain. Differences at the territorial level should not be underestimated, because they might indicate different

Figure 9
TOTAL FERTILITY RATES AT EUROPEAN, NATIONAL AND REGIONAL LEVEL, 1960–90

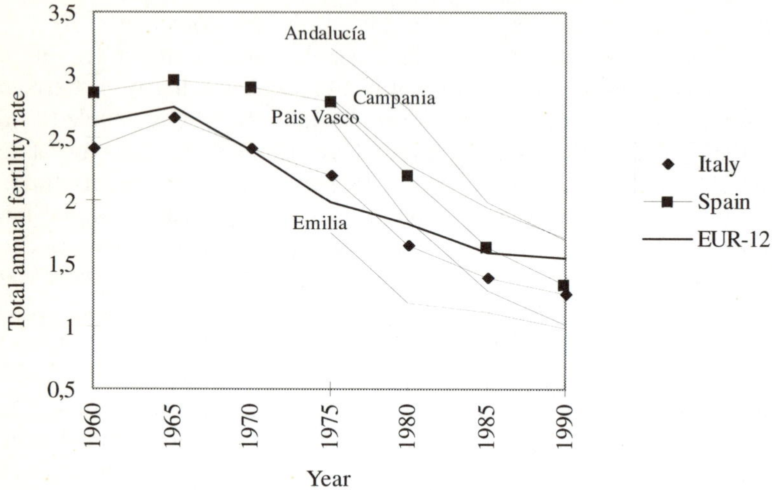

Source: Eurostat (1992b), table E-9. Delgado Pérez/Livi-Bacci (1992).

family cultures within a country (Saraceno, 1994).

The second characteristic which distinguishes Italy and Spain from the rest of Europe is that the decline in marital fertility has not been accompanied by a substantial increase in births outside of marriage.

The proportion of births outside marriage has been increasing in all European countries, rising from an average of 5.1 per 100 live births in 1960 to 23 per cent by the mid-1990s. In Italy that figure was 7.7 per cent in 1995 and 10.8 per cent in Spain (Eurostat, 1995). In other words, the birth of children in southern European countries is more strongly related to marriage than in other European member states, and fertility patterns should be viewed in relation to marriage patterns. Clearly, the number of births outside of marriage is low in these two countries, because this phenomenon is associated with high unmarried cohabitation and a high divorce rate.

Explaining the rapid decrease of fertility and the comparatively low fertility rates in Italy and Spain is not an easy task. It is especially puzzling because both countries are Catholic and are considered to be comparatively highly child-oriented (Jurado and Naldini, 1996). It is also difficult to explain the decline in fertility as the result of modernization or by using the explanations given for other western European countries (higher rates of

Figure 10
BIRTHS OUTSIDE MARRIAGE, 1960–95

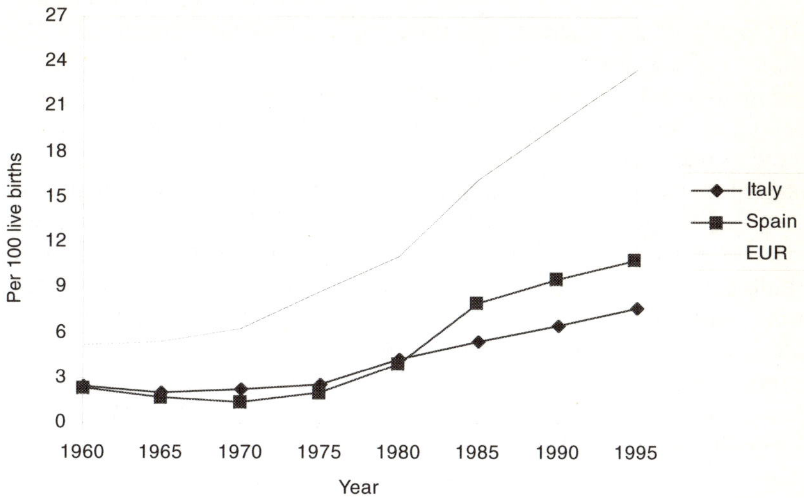

Source: Eurostat (1995), table E-4.

female employment, higher education levels among women, the sexual revolution and secularization). The reduction of fertility rates in these two countries occurred and accelerated in the presence of the following factors: (1) a lower rate of women in the labour market in comparison with other European countries; (2) a gender gap in education that remained stable until the 1970s (later than in other countries); (3) legislation that up to 1975 in Italy and 1978 in Spain prohibited abortion as well as the dissemination of both knowledge and the use of contraception (even after abortion and contraception were legalized, only an incomplete contraceptive revolution took place); and (4) the strong influence of the Catholic Church on political, cultural and daily life, although to a different degree in both countries (Saraceno, 1994).

Some Spanish and Italian analysts (Delgado and Livi-Bacci, 1992) suggest that the major explanatory factor in the drop in fertility is the late and massive entry of Italian and Spanish women into the formal labour force since the 1970s. These changes occurred so fast that social norms, not to mention child care, family allowances, labour legislation, housing and other social services, were not able to keep pace. It seems significant that Italy and Spain have some of the least generous child benefit packages of all European countries (Bradshaw and Ditch, 1993). If this analysis is correct, childbearing presents greater opportunity costs to Italian and Spanish women than to women in other west European countries. In these countries women's participation in the non-agricultural sector of the labour force has a longer tradition and different institutional characteristics (i.e. diffusion of part-time).

Employment structures only offer an incomplete explanation for the declining fertility rates, however; the delay of marriage surely influences women's reproductive behaviour in Spain and Italy. Thus, the causes for the delay of marriage (high youth unemployment rates, long education phase, housing shortages, etc.) may influence the fertility level in an indirect way. However, in explaining the reduction of fertility rates in these two countries, I think one has to take account of new family strategies, as well as changes in the experience and models of being mothers and fathers. In Italy, the reduction of fertility and the extended delay of the first birth after marriage do not mean that women are renouncing maternity. In this country the number of couples with at least one child remains higher than in other European countries. Though there is great regional variation in both countries, the fertility decline in Spain and Italy results mainly from couples' decisions to have only one child instead of two or more. Italy and Spain's child-orientation no longer means having many children, and couples can fulfil their desire to be a parent with only one child. Furthermore, in a context characterized by social and economic constraints, as well as deep changes in gender relationships, young parents are more concerned with developing the quality of their relationship with each child rather than with having large numbers of children.

Family Dissolution Phase

Italy and Spain introduced divorce laws later than did other European countries, in 1970 and 1981 respectively. This delayed introduction seems to be characteristic of Catholic European countries in general (e.g. Ireland). The demographic characteristics of people who get divorced in countries with a late introduction of divorce law are very different from those whose countries introduced divorce at the beginning of liberalization. Divorcing couples in late-comer countries may have already been married or *de facto* separated for a long time, and thus are typically older and less likely to have young children (Goode, 1993). Although divorce in Italy seems to be somewhat distinct from other industrialized countries, Barbagli (1990) demonstrates that the same general tendencies hold true in this country as well.

First, divorce in Italy (and Spain) is not a substitute for a separation, but is simply added to a separation as a legal step. In these two countries spouses must demonstrate that they have been legally separated for a minimum period before they can obtain a divorce. The divorce rates in these two countries are still much lower than those in other western countries. In 1989 the number of divorces per 1,000 people was 0.6 for Spain, 0.5 for Italy and 1.6 in the European Twelve (Eurostat,1992b).[4] However, a correct comparison of conjugal instability in Italy and Spain and the other European countries should include not only the number of divorces but also the number of legal separations. This number in Italy and Spain has not been so different from the number of divorces in other countries. Finally, many couples remain at the stage of separation. In Spain a survey on marital dissolution shows that of all couples in 1990, 0.9 per cent broke down because of a divorce, 2 per cent because of legal separation and 3.3 per cent because of *de facto* separation. This means that until 1990 a lot of Spanish people preferred extra-legal solutions to their marital problems (Alberdi, Flaquer and de Ussel, 1994). In Italy in 1991 the number of legal separations was 44,920, while the number of divorces was only 27,350 (Golini and Menniti, 1994). Therefore, if we used conjugal instability rates instead of divorce rates for our comparisons, the divorce gap between Italy and Spain and the rest of Europe would be considerably reduced.

A second important difference between Italy and Spain and the western European countries, in part a consequence of the complexity of Spanish and Italian divorce law, can be found in the ages of Italian and Spanish divorcees. They are on average much older than in other countries. In Spain in 1991 more than 60 per cent of the divorced or legally separated were 35 years and older and 26.9 per cent were between 45–54 years old (Valero, 1992).

Third, in Italy the divorce rate is still higher among the upper social classes, although Barbagli asserts that this difference will decline over time (Barbagli, 1990). Borrajo Iniesta (1990) has also found a positive correlation between frequency of divorce and social class for the city of Madrid (Iglesias de Ussel, 1994). In addition, in Italy the evolution of the phenomenon has, and always has had, different values between the north and the south. As Barbagli notes, in more northern cities the legal separation rates are the same as those of Sweden.

A fourth difference Barbagli reports is a definite correlation between number of children and reluctance to divorce, that is, couples with more children divorce less often than couples with fewer children. This correlation is less prevalent in other Western countries (Barbagli, 1990). The findings for Spain are not so clear. A study on divorced and separated women in the city of Madrid for 1981–84 shows that this correlation does not exist if researchers control for the time of marriage (Borrajo Iniesta, 1987). Borrajo Iniesta corroborates these findings in a study using national survey data from 1981 to 1986. He found that legally separated or divorced couples did not have fewer children than married couples, with the exception of couples married between two and eight years. In this case separated or divorced people had fewer children (Borrajo Iniesta, 1990).

To conclude, divorce was introduced later and in a different manner in the southern Catholic countries and this has influenced couples' behaviour. Since divorce requires more time and money, most couples only ask for legal separation. Divorce is most common among older people and the upper classes in the northern urban areas. Children still seem to be a barrier to divorce.

Intergenerational Ties and Kinship Solidarity

The nuclearization of family households does not necessarily mean greater isolation from extended kin networks. On the contrary, during the last decades national and cross-national research has shown that support within kin networks is still very important and that the modern nuclear family is actively involved in kinship and social networks (Bruckner, Knaup and Müller, 1993; Höllinger and Haller, 1990; ISTAT, 1985; De Miguel, 1992). This section tries to answer the following questions: how strong are kin relationships in Italy? are they similar to the Spanish ones? are they stronger or weaker than those in other western countries for which there are comparable data?

The spatial distance between kin, the frequency of contact, and the functions of these kin relations, especially between adult children and their parents, are the three main dimensions of kinship social network analysis. It has been more difficult to include measures of the affective meanings of such relations, and to make conclusions on their emotional quality or intensity. This is especially true in cross-national comparisons, where different outcomes can be the result of sociocultural differences in the interpretation of survey instruments.[5]

Höllinger and Haller (1990) present results from the International Social Survey Programme (ISSP) survey, showing that kin relations, especially intergenerational family relationships, such as those between parents and children, are still very important across all countries. However, they show that Americans and Australians have loosened kin ties further than have the Britons, Germans and Austrians. Hungarians and Italians do still maintain very close kin relations. The authors point to levels of economic development, urbanization, mobility, sociocultural family patterns, and their mutual interaction as the explanatory variables for international variation in the intensity of kin social networks. They also discuss preindustrial traditions of close kin relations in southern and eastern Europe versus loosened kin ties in north-western Europe. Finally, they also point out the higher geographic mobility among Americans and Australians, and the individualistic lifestyle promoted in Great Britain. Italians are most likely to settle in the immediate neighbourhood of their parents and have the most frequent contact with their kin. Less than 10 per cent of Italians and Hungarians indicate that they meet their parents only a few times a year.

As can be noted in Table 5, both the kin and friendship networks in Italy are wider than in Germany. Relatives in Italy live closer together than in Germany and visit each other more often. Italian social networks, both kin and friendship relationships, are characterized by a strong 'intensity' and 'localism'. Unlike the Germans, for whom kin networks are namely relationships with the mother and father, the Italians extend kin relationships beyond the intergenerational family solidarity, towards siblings and other relatives.

The studies based on 1986 ISSP data (Bruckner, Knaup and Müller, 1993) show that in general the importance of friends in social support networks is inversely proportional to the importance of kin relations. This does not seem to hold true for Italy. Italians expect social support from both relatives and friends. The more intense relations of Italians with their relatives also results in a greater importance of relatives for social and economic support.[6] For help, Italians mostly ask relatives (mother, father, but also sisters and brothers). They are less likely than the Germans to name their partner as the 'Number One Helper'. This difference may imply that the set of persons Italians can turn to is larger due to closer kin ties and the fact that they feel less obliged to turn to their partner (Höllinger and Haller, 1990; Bruckner, Knaup and Müller, 1993).

Table 5
PEOPLE VISITING THEIR RELATIVES: WEST GERMANY AND ITALY IN 1986

Category of relatives not living in the same household	**West Germany**	**Italy**
Mother	12.8	15.7
Father	11.1	14.0
Daughter	17.6	18.1
Son	14.5	13.5
Sister	5.5	14.9
Brother	4.9	16.0
Other relative	9.5	26.0

Source: ISSP (1986), pp. 32–48, 61, 63.

Is Spain similar to Italy? A Eurobarometer survey in 1992 on older people and their attitudes in EU countries gives some clues to this. Elderly people (over 60 years) were asked, 'How often do you see your family?' The results support the previous analysis on social networks in Italy and Germany (see Table 5), and I hypothesize that Italy and Spain have a similar pattern of kin relations.

Figure 11
DAILY VISITS OF THE ELDERLY, 1993

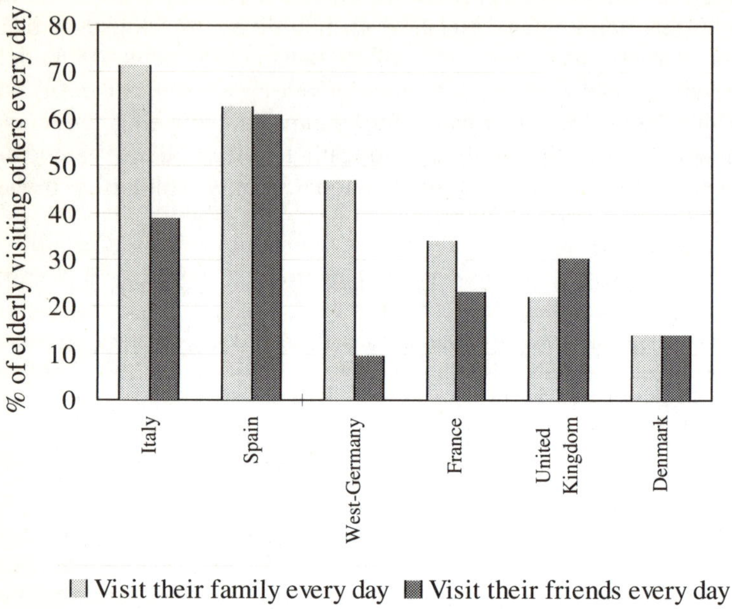

Source: Eurobarometer (1992), 37.2.

However, it is worth remembering that these data only present national averages and that they do not provide information on geographical or social differences within the country. Analysis of the Italian survey data (ISTAT, 1985) on household structure and organization and the recent *Multiscope* survey (ISTAT, 1993) reveal many differences in the functioning and exchange of resources across kin networks throughout the country. In the north families give help – to kin and non-kin – more than they receive it, indicating a situation where the need for support is not too serious or is met through social services. By contrast, in the south families receive more help than they give, and the amount of aid received is on average higher than in the northern regions (Sabbadini, 1994; Saraceno, 1998). In addition, the ISTAT survey revealed important differences between genders, social classes, metropolitan areas and medium and/or small towns, and between urban and rural areas (Sgritta, 1986). Finally, Italian studies confirm previous research on social networks, which revealed the feminization of kin relations and the central role played by women, who give more help to kin than men (Facchini, 1997; Sabbadini, 1994).

Many Italian studies show that grandmothers are the main source of child care for young families, especially in cases where the mother works. This holds across regions, even in places such as Emilia Romagna, where the system of public day-care services for the very young is quite good (Musatti, 1992). Older generations also receive support: elderly who live in one-person households are rarely isolated. They live within a network of exchange and they receive help when they are disabled or ill. Daughters and daughters-in-law provide most care for elderly family members. The 'woman sandwich' is the name given to women in the central age (40–60) who are divided between the care of children and the care of the elderly.

Studies on social and kin networks in Spain have been very limited. Available research, however, seems to indicate a high number of contacts and exchange between close kin, especially across generations (Alberdi, 1994; De Miguel, 1992). According to a survey carried out in 1991–92 by Cires, one-third of all Spaniards reported daily or almost daily visits to a kin member living in a different household (Cires, 1993; Iglesias de Ussel, 1994). A study on elderly life in Bilbao, which aimed to describe the family situation and networks of the oldest in that town, also supports this conclusion. The author argues that Litwak's category of 'extended family cohesion', in which the nuclear family is not isolated but well integrated into kinship networks, best describes the life of the elderly in Bilbao (Bazo, 1990; Litwak, 1960).

The level of economic development of a country as indicated by industrialization, urbanization and mobility can only partially explain the persistent importance of primary relations in some countries. Höllinger and Haller (1990) have demonstrated that although Italy has a higher degree of urbanization than Austria, kin relationships are much more important in the former country. In addition, in Italy close kin relations are quite common both in rural and urban areas and even the most industrialized regions of the country give little indication of the decreasing importance of kin relations.

Data show that family members in both Italy and Spain are more strongly connected to each other, and that kin networks are more important than in other countries. The greater length of time that different generations live together, coupled with the relatively higher institutionalization of marriage (as measured by low rates of cohabitation, out of wedlock births and divorce), might explain the higher density of contact within and between households. But how can we explain the high level of support and solidarity within networks in these countries? Is it due to the high density of contacts or to other socio-economic characteristics? To what extent can the tradition of the extended and/or multiple family structure and the different distribution of the agricultural population in the past in some areas and regions explain longer cohabitation across generations?

The literature on this issue, although rich for some countries (Laslett, 1972), gives no comprehensive picture of family types and of kinship systems in the past. Especially among Mediterranean countries, family structures and marriage patterns are not easily reduced to one model among and within countries (Barbagli and Kertzer, 1992; Rowland, 1987). For instance, for Italy it has been found that the predominance of the rural population in some regions and the form of prevalent agricultural contracts (i.e. sharecropping and *boaria*) account for the great regional and provincial variations in family structures, as can be seen by the 1951 census. Nuclear families were much more widespread in the south, while the incidence of extended and multiple families was higher than the national average in the central regions, particularly in Tuscany, Umbria and Emilia-Romagna.

In addition, a high diversity in family formation between urban and rural areas as well as between the south and the centre-north has persisted since the period before industrialization (Barbagli, 1988; Barbagli, 1996; Saraceno, 1997). Despite these important regional variations in family structure and family formation, as well as the differences in the functioning and types of exchange in kin networks, intergenerational ties and kin solidarity are still important throughout the entire country. This shows, therefore, that it is wrong to assume that the advent of the nuclear family went hand-in-hand with a loss of intergenerational ties and kin solidarity, and that it meant the loss of its welfare functions.

The religious denomination and the strength of the Church is another indicator of the cultural traditions concerning family solidarity in these two countries. As noted above in the discussions on state–Church relations during the authoritarian periods and the Church's influence on family law, especially the late approbation of divorce law, the Catholic Church in Spain and Italy has traditionally had a great influence on family matters. As has been shown before, the situation of the labour market in these two countries is crucial for understanding family changes. In Italy, regional variance analysis reveals relationships between the level of female employment and fertility and divorce rates (Barbagli, 1990). This points to the fact that certain characteristics of the labour market that are more accentuated in certain areas of the country might also affect gender relations. For example, in the southern regions the low rate of female employment and the high presence of women in the informal sector and among precarious jobs favours an asymmetric relation between genders in the family; southern Italian women may therefore find it more difficult to seek a divorce.

A factor that should not be underestimated in explaining cross-national variations in kin relations is the development of formal services, public or private, related to caring work, such as child care and nursing or rest homes for the aged. The development of these services, and the degree to which their purposes are defined as being primary or 'subsidiary' *vis-à-vis* the family, affect intergenerational living arrangements and intergenerational and kinship support. Moreover, the nature and the characteristics of social policies and law and the definition of family and state responsibilities also influence the nature and the intensity of intergenerational ties and kin solidarity. Formal, institutional support is clearly complementary to informal, family-provided support. Furthermore, the less formal supports that are developed, the more important the informal supports become. In northern Italy, where social services are more developed and labour resources (i.e. employment) are greater, people give more help than they receive; the opposite is true in southern Italy. The territorial differences may point to a higher dependency of southern families on the family and kin solidarity than northern ones. However, the differences between the size of the town in which people are living and the social classes to which they belong are also very important in shaping different models of solidarity within a country. In the next chapter I will illustrate how the permanence and strength of primary social networks in southern European countries have been taken for granted and reinforced by the characteristics of certain social policies and laws and by the development of a 'dualistic' protection system.

NOTES

1. Some of the ideas contained in the next sections were developed while I was working with Teresa Jurado Guerrero. For the result of that work, see Jurado and Naldini (1996).
2. So far national indicators are the only way to make comparisons between countries. It should be mentioned that while these represent the 'average' of sub-populations, within every country there are many variations across territorial and social levels.
3. Some analysts have suggested that this number may be largely underestimated (Sabbadini, 1991; Saraceno, 1998).
4. This indicator is very rough, but it is the only one available for Spain. The better indicator for international comparisons, i.e. divorces per 1,000 existing marriages, is not available in Spain for non-census years and not yet for 1990 (cf. EUROSTAT, 1992b).
5. It should be noted that comparative data on family and kinship networks are still lacking. The only comparative study on social networks available is the International Social Survey Programme on Social Relations and Social Support (ISSP; 1986), which includes seven countries: Australia, Austria, Britain, West Germany, Hungary, Italy and the USA. Unfortunately the survey does not include Spain. Thus, I will refer to the Italian and German cases as representative of two different clusters of European countries.
6. In the ISSP (1986), people were asked about who helps them in their houses and gardens, in the case of flu, depression, marital crisis, and from whom they request financial help and advice about important changes in life.

Chapter 7

SOCIAL POLICY AND SOCIAL SERVICES FOR THE FAMILY

The purpose of this chapter is to find out whether and to what extent the new democratic regimes changed past policy relating to the issue of family responsibilities. The aim is to evaluate which family models – the male breadwinner or the family/kinship solidarity model – have been supported by the Italian and Spanish welfare states. The analysis intends to show that the Italian and Spanish welfare states supported the male breadwinner family model[1] only to a limited extent. In fact, in these two countries a low level of economic support for families with children is found. Furthermore, the focus of state intervention has continued to be, as it was during the past dictatorial regimes, not only the nuclear family but also the kin. In particular, the emphasis in this chapter is placed on four policies: survivors' pensions; family allowances; tax relief; and maternity and child care programmes.

The Italian Case

After the downfall of fascism, Italy abolished almost all fascist-era family benefits, particularly those with pro-natalist objectives. The only exception was family allowances, which, though maintained, were soon stripped of their pro-natalist aims. Among fiscal policies, tax allowances for family dependants remained but soon lost any demographic intent. Real tax reform was not implemented until the beginning of the 1970s.

In 1945 survivors' pensions were introduced; maternity leave was reformed in 1950 and once again in 1971. In 1968, in the framework of educational policies, Italy instituted the *scuola materna* for children between 3 and 6 years of age. The sectors with the most long-lasting policy legacies were social assistance and child care facilities. In particular, provisions for children under 3 continued to be run by the Opera Nazionale Maternità e Infanzia (ONMI) until the new law of 1971. Finally, it is worth noting that in most recent times a new interest in families with children is observable in Italy. On the one hand, new types of means-tested family benefits have been introduced (e.g. household allowances for families with three children, maternity allowances for women not insured). These new benefits have resulted in a fragmented social policy system, which is mainly aimed at supporting low-income families with children under 18. On the other hand, in what seems like a new wave of interest in innovative approaches to family policy, combined with the push of EU regulations, new reforms have come into force (a new law supporting working mothers and parental responsibilities has been introduced very recently).

Survivors' Pensions

Italian survivors' pensions are granted mainly to spouses, although they are also granted to orphans and to a lesser degree, to parents since 1952 and to siblings since 1965. The foundations of Italian survivors' pensions are found in the 1939 law. However, the scheme came into force only in 1945.[2] The survivors' pension provisions were originally framed as a means of assuring dependent family members' economic subsistence after the (male) breadwinner's death. Their purpose was clearly one of assistance and of support for the male breadwinner. The idea of the male breadwinner family model was reinforced by the maintenance, up until the mid-1970s, of two gender-discriminatory elements embedded in survivors' pension law. The first concerned the male and female orphans of the breadwinner. Until 1975, the survivors' pension could be paid to married sons, but was not paid to married daughters.[3] The second discriminated against survivors of deceased female workers. Pensions of deceased wives were paid out to the surviving family members only in cases where the husband was recognized as disabled or unable to work.

The Constitutional Court (Corte Costituzionale) declared this rule unconstitutional, since it violated the principle of equal treatment between men and women.[4] Since then, the survivors' pension is paid to a surviving spouse independently of his/her earnings or pension, and children may benefit from their deceased mother's pension independently from their father's income. A degree of income testing in survivors' pensions has been introduced with the 1995 pension reform.

Since 1952 the entitlement to survivors' pensions has been extended to elderly parents, and since 1965, to siblings. The extension of entitlement to kin is illustrative of the importance attributed by Italian social policies to family/kinship ties. The supplementary nature of this provision is evident in that the parents' entitlement to survivors' pensions (as well as that of siblings) was given only under very strict conditions (i.e. neither the deceased's spouse nor the orphans must be entitled to it; parents must be aged over 65; parents must not have their own pension). In addition, each parents' survivor's pension amounted to only 15 per cent of the deceased workers' pension.

In 1965 a new law reformed the survivors' pension programme and enlarged kin members' entitlement to siblings' pensions. The reform of 1965 added a new condition for parents' pension rights: they had to have been maintained by the deceased pensioner before his or her death. If neither of the parents were entitled to the allowance, it could be granted to unmarried siblings and/or orphans over the age of 18, provided they were permanently unable to work and had been economically dependent on the deceased pensioner.

The first survivors' pension scheme introduced during the democratic regime was the same one that had been planned during the fascist period. While the range of those entitled was expanded during the reforms of the first decades of the democracy, the institutional features remained the same.

Family Allowances

In Italy family allowances were and still are connected with gender, intergenerational and kinship relations. After the fall of the fascist regime, family allowances lost their pro-natalist objectives; after 1944, birth order and the number of children were no longer used to determine the amount of the allowance. Notwithstanding the abolishment of a few articles of the previous fascist law, the policy confirmed entirely the second goal of the past regime: the acknowledgement of male workers' economic responsibility for their children, wives, and other dependent family members. Eligibility was granted to most wage workers (i.e. the head of the family, legally defined as the husband until 1975) employed in the private sector with dependent family members. In this regard it is necessary to remember that family allowances were created as social security and contributory benefits.

In 1945 a similar measure was introduced for public employees in the form of a supplementary wage. That same year a contribution ceiling (*massimale di contribuzione*) was introduced, exempting employers from paying social contributions towards family allowances on wages over a defined threshold. During the 1950s and 1960s the only relevant change in the Italian scheme was the gradual extension of the benefit to other social groups (unemployed and pensioners) and to those working in the agricultural sector (self-employed farmers and sharecroppers). This extension of the coverage of family allowances resulted in a growth in the number of beneficiaries until 1974, as well as of expenditure until 1970 (Franco, 1993). Although the number of beneficiaries increased, inflation progressively eroded the value of the allowance (see Figure 12). Since then, the importance of family allowances has drastically dropped in terms of their real value for recipients, as well as in terms of their share of total expenditures as a percentage of GDP, and of total social expenditures.

Figure 12
INDICATORS OF FAMILY BENEFITS EXPENDITURES, 1960–87

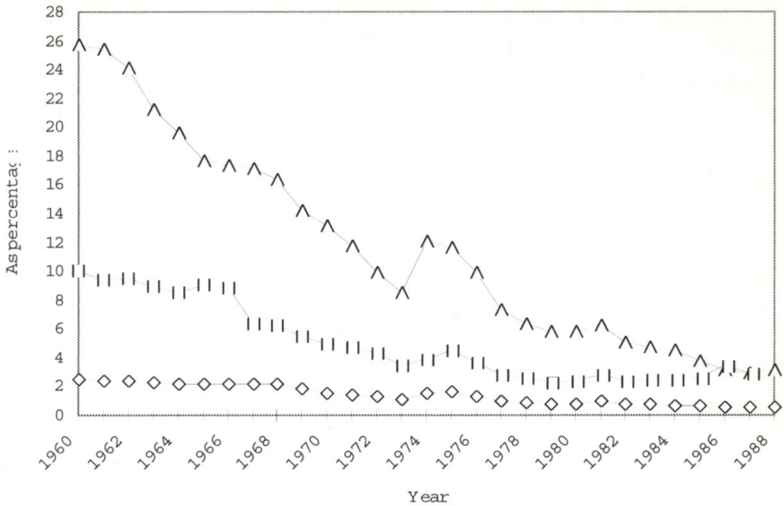

Key: —◇— Expenditure as % GDP

—△— Expenditure % of total social expenditure

—□— Value allowance as % GDP per capita

Sources: Franco and Sartor (1990); Franco (1993); ISTAT (1983a); ISTAT (1983b).

At the same time, since family allowances were financed by employers' contributions based on total wages (with no upper limit after 1974), family allowance funds have been growing. Part of the family allowance fund surplus has been used to make up deficits in other sectors of the National Institute of Social Protection (Istituto Nazionale della Previdenza Sociale) (Gorrieri, 1979).

Reforms in the 1980s Italy had to wait until 1988 for a reform of family allowances. Indeed, the Italian reform of family allowances began in 1983 with the introduction of a 'means-tested supplement' (*assegni familiari integrativi*) for low-income families with children under the age of 18. The broader reform of family allowances occurred five years later. The 1988 reform established the 'household allowance' (*assegno per il nucleo familiare*), as it is now called, a complete means-tested measure for low-income households. However, this benefit is still framed as a social security, contributory benefit, and is therefore limited to dependent workers and employees. Income thresholds are defined for different family sizes, which entitles recipients to allowances of varying amounts. Thus, the number of family dependants or children no longer determines entitlement to the allowance. Rather, the amount of total family income with respect to the total number of household members is relevant.

The worker, his or her spouse (unless legally separated), and children under age 18 or disabled children are included when calculating the size of the household. Siblings and grandchildren under the age of 18 or disabled grandchildren may be included if they are orphans and not entitled to a survivors' pension. Finally, the system acknowledges the higher expenses of one-parent families and families including invalids or disabled people by raising their income thresholds.

Over time, family allowances have lost their purchasing power, because while income thresholds are indexed to the cost of living (but not to trends in wages), the corresponding allowances have remained the same since 1988 (CIPE, 1995). It was only since the mid-1990s that attention to families with minor children was evident. In fact, since 1994 a set of additional measures introduced by financial budget laws (*leggi finanziarie*) has increased the amount of household benefit and adjusted the family income threshold for families with children under 18 years.

The 1988 reform has produced a policy hybrid: while in principle it is a social security, contributory benefit – and therefore limited to dependent workers and employees – it is not only means-tested, but largely paid for directly from the state budget. In order to alleviate poverty, which is especially high in Italy among larger families, a new means-tested benefit was introduced in 1999. Since then households with at least three children under the age of 18, and with an annual household income of under 18,595 E (calculated with ISE, *Indicatore della Situazione Economica*, Indicator of Economic Condition) and belonging to a household of five members are entitled to a monthly household allowance of 103 E for 13 months. This new benefit has to be seen as a measure against poverty rather than as a measure aimed to support the family.

Taxation Policies: Tax Unit and Tax Relief for Dependent Family Members

Taxation policies relate to the family in two different ways. First, the tax system's treatment of husbands' and wives' respective incomes may indirectly encourage or discourage women's employment. Therefore, fiscal policies based on the family as a unit of taxation point to a specific institutional definition of 'family'. For example, under a joint taxation system and a progressive marginal tax rate, a couple is penalized when the wife works.

Second, income tax systems can make adjustments to the tax paid by families with children by offering tax credits or allowances for children,[5] or by adjusting the rate of tax for the presence of children and/or other dependent members in the household (i.e., splitting). Thus, tax relief is similar to family allowances in supporting families with dependent family members.

In the new Italian Republic, tax allowances for families with children which seemed to have demographic purposes were abolished by 1946 (Franco and Sartor, 1990). However, the core of the taxation system created during the regime was maintained in Italy up to the tax reform of 1971. This reform[6] abolished the fragmented system of fiscal policies and introduced the personal income tax (IRPEF). The family continued to be considered as the unit of taxation (through the legal concept of the family head, which was the husband up to 1975), so that joint taxation was mandatory for married couples. The discrimination between equal family incomes, but produced by a different number of earners, was achieved through tax relief for each separate earner (Franco and Sartor, 1990).

As far as tax relief for dependent family members is concerned, the most important innovation of the reform of 1971 was the replacement of tax allowances (*deduzioni dal reddito*) with tax credits (*detrazione d'imposta*). Tax credits were fixed at a higher amount for dependent spouses than for dependent children to compensate for the cumulative spouses' income. Children were defined as dependent if they were under the age of 18; under the age of 21 if in school or an apprentice; under the age of 26 if enrolled in university; or if they were permanently unable to work. In addition, tax credits for other members of the family (as during fascism) were maintained. In this regard, a tax credit was granted for other dependants, provided their income did not exceed a given threshold. 'Other dependent family members' (*altri familiari a carico*) referred to a cohabiting person with an income below the given threshold, who would be entitled to a maintenance allowance under civil law (kin maintenance law, art. 433 civil code). Adult children over the ages of 18 or 26 who had low incomes were also treated as dependants.

As a consequence of the 1975 family reform, and of the abolition of the legal concept of 'family head', the Constitutional Court declared the principle of joint taxation for spouses to be unconstitutional. Since then, Italy applies the principle of individual taxation, that is, spouses are now entitled to separate assessment.

An interesting innovation introduced by tax reform law which anticipated the decision of the court, is the law of 1975 on family firms (*impresa familiare*), which established that the revenues produced by a family firm can be split up between the entrepreneur and the other family 'collaborators' (*collaboratori familiari*). Family collaborators include the spouse, kin up to the third degree, and kin-in-law up to the second degree (art. 230 bis civil code). In this case, family kinship ties are encouraged and supported explicitly by Italian fiscal law.

The structure of IRPEF was modified in 1983 and again in 1986, reducing the tax brackets from 32 to 9, and revising rates, ceilings and tax credits. Further adjustments were made, mostly in yearly fiscal laws, but they did not change the overall tax structure. An estimation of the weight of tax credits for dependent family members over time is not an easy task. The composition and size of households is taken into account in very different ways: tax credits, tax allowances, *quotient familial* and splitting. Moreover, provided that income tax is assessed on a progressive basis, any instrument which aims to achieve horizontal equity (between families of different sizes and composition) cannot be evaluated without taking into account that income tax is mainly an instrument designed to achieve redistributive effects (between families with different incomes).

According to Franco and Sartor (1990), it has to be remembered that any evaluation of the amount of tax credits cannot be separated from an analysis of tax schedules and gross rates of the same period. For this reason the

Figure 13
TAX CREDITS FOR DEPENDENT FAMILY MEMBERS, 1974–88

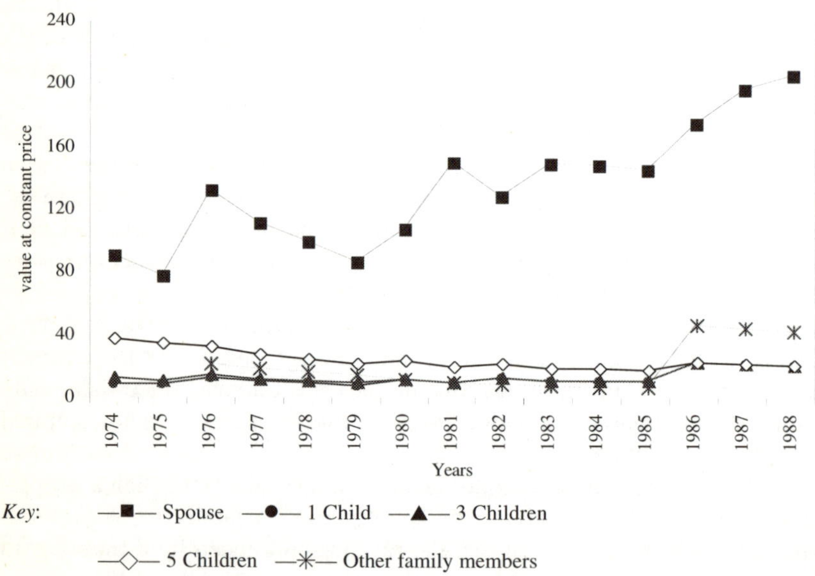

Source: Franco and Sartor (1990), table 11, p. 120.

analysis of the evolution of tax credits (at constant prices) should be read within the periods 1974–82, 1983–85 and 1986–88, because within a given period tax schedules and rates are held constant.

During the period 1974–82 the evolution of tax credits has decreased, apart from re-evaluations made in 1976, 1980 and 1981. In the following period, 1983–85, tax credits were indexed to consumer prices. From 1986 to 1988, consumer price indexing was removed. It is worth observing that since the mid-1980s the value of tax credits for other dependent family members has become twice that of children. Finally, it has to be said that in Italy public support for families with children, both in terms of tax credits and in terms of cash benefits, has not only been declining over time but also in comparison with other European countries (cf. below, appendices 4 and 5).

In 1998 the system of tax credits for children and for other dependent family members was reformed in Italy. The new tax credit reform seems to recognize the long dependency of Italian youths on their family. The distinction between tax credits for children under age and other economically dependent members (adult children over the fixed age limit are also treated as dependants) was abolished. The amount of tax credits for dependent adult children was increased to the rate for children, giving each taxpayer a tax credit for each dependent child (or for any other family dependent member), with no age limit.

Maternity Leave Reforms and Child Care Provisions

Maternity (and parental) leave policies may be regarded as the way in which states assign women's appropriate role as mother, worker, or both. Public child care policies may be regarded as the means by which the care of children is acknowledged by the state, not only as a family (woman's) responsibility, but also as a public responsibility. Child care policies, in turn, influence and interact with women's employment in different ways. However, public child care facilities may have been designed to achieve different goals: either to permit women to combine paid work and motherhood, or to provide better conditions for the education and care of young children.

Maternity leave reforms Soon after the Second World War Italian trade unions and the Unione Donne Italiane (UDI) (the main Italian feminist association) campaigned actively for a new law for motherhood, which was primarily intended to improve the health of working women. In the Italian parliament the final bill was presented and promoted by the trade unions. The proposal, which was rather advanced for the period, was translated into the less progressive 1951 law (no. 860, *'tutela fisica ed economica delle lavoratrici madri'*). This law introduced five main changes. First of all, the compulsory maternity leave period was extended from a total of ten weeks to three months before and two months after confinement. Next, women on maternity leave were entitled to 80 per cent of their normal pay. Third, mothers could not be dismissed from employment until after the baby was 1 year old. Fourth, new mothers had the right to daily time-out for breast-feeding during the baby's first year. Finally, employers had to provide facilities for mothers to feed their babies in the workplace. Employers who employed at least 30 married women were asked to organize and/or contribute to the creation of nursery rooms. Despite the fact that women were often employed in small enterprises, the law demanded no financial responsibilities from the employer if the company employed fewer than 30 married women.[7]

The new democratic government's maternity law was in many respects more continuous than discontinuous with prior legislation. First of all, while paid maternity leave was extended for the first time to agricultural workers, home-workers (*lavoratrici a domicilio*), mostly women, were still excluded, as was the case during fascism. Second, employers were still responsible for the custody of the babies of their employed working mothers. Women's employment continued to be considered more costly than men's. Third, the state's commitment to child care provision remained almost unchanged and was still managed by ONMI. Finally, while the new maternity law guaranteed the welfare of working mothers and their newborn children through compulsory maternity leave, it did little to change women's social conditions or their traditional role in the family.

During the 1950s and the 1960s Italy failed to develop policies and/or child care programmes that would support female labour market participation and assist women in combining work outside the home with motherhood (Ballestrero, 1979). Therefore, traditional beliefs that the care of children is a family matter, and that the care of the family is mainly a woman's duty, remained unchallenged.

In the 1970s the emergence of new social movements, namely the women's movement, trade unions and social rights movements, triggered a great deal of reform and welfare state expansion in Italy. In 1971, within this context of renewal and discourse about social rights, Italy also enacted both maternity law reform in the law on working mothers (law no. 1204, '*sulle lavoratrici madri*') and a day-care law (law no. 1044, on '*asili nido*' or '*nidi*').

Law 1204 on working mothers extended the right of maternity benefits to home-workers and domestic workers and, after 1977, to working mothers with adopted or foster children. Moreover, it extended maternity leave from two months to three months before the childbirth if the expectant mother's work activity was particularly hazardous. It also established that the mother could not be employed in dangerous and/or heavy activities for up to seven months after childbirth.

In comparison with the maternity law of 1951, this reform introduced two important innovations. First, it introduced the option of extending maternity leave for an additional six months at 30 per cent of the mother's normal wages. The second innovation was the introduction of unpaid leave for caring. Mothers were granted the right to take unpaid leave to care for sick children under the age of 3. After 1977 these options were also available to fathers, and the title of the programme was subsequently changed to parental leave.

For almost three decades maternity and parental leave legislation, which it has to be said was one of the most advanced at that time in comparison with other European legislation, remained unchanged. It is only in the past two years that reforms have been introduced. Since 1999, in effect, mothers who are not receiving maternity insurance benefits are entitled to a means-tested maternity allowance benefit, for a maximum of five months.

Second, a reform of the Italian parental leave system in March 2000 introduced greater flexibility and a longer leave period. The new law[8] is intended to enable parents to combine family and work life. It explicitly recognizes the role of the father and his right to parental leave. Furthermore, it introduces the principle of shared responsibilities after childbirth and during the child's upbringing. Now both working mothers and working fathers are entitled to an 'optional' parental leave of six months each for a total of ten months. However, the father has the right to a 'parental leave prize' of one month if he takes at least three months of parental leave (for a maximum of 11 months in total). Under the new law, the parental leave option is extended to the child's first eight years (it previously covered only the first). Small firms (with fewer than 20 employees) that need to temporarily replace an employee on parental leave are entitled to a reduction in their social security contributions for the newly hired worker. Firms can also apply for a grant to support the introduction of flexible working hours, telecommuting, and work at home for employees returning from their parental leave.

The current system of child care services Since 1971 Italian public child care services, like those of many other European countries, have been composed of two systems, according to the age of children: day-care services (*asili nido*) for children under 3, and kindergartens (*scuole materne*, now named *scuole per l'infanzia*) for children between 3 and 5. These services are publicly funded and run by the state and/or local authorities.

The Italian kindergarten system, '*scuola per l'infanzia*', is a non-compulsory educational pre-school service and as such forms part of the national educational policy. By the time the 1968 law was enacted, the state had assumed direct responsibility, alongside local authorities, for the funding and organization of these pre-schools. Local governments and the state, therefore, have become the two main providers (in terms of funding and organizing) of this service. The state, however, opted for subsidizing licensed, private non-profit kindergartens (mostly run by religious organizations) in locations where public and subsidized private facilities did not meet the demand for services. As a result of direct state involvement, by the early 1970s around 60 per cent of all children between the ages of 3 and 5 were attending kindergarten. As Table 6 shows, the figure rose to 70 per cent by the mid-1970s and now reaches almost 95 per cent.

Table 6
CHILDREN BETWEEN 3 AND 5 YEARS ENROLLED IN *SCUOLA PER L'INFANZIA*, BY GEOGRAPHICAL AREA IN ITALY

Date	North-west	North-east	Centre	South	Italy
1976–77	70.8	74.0	70.2	65.1	69.0
1977–78	73.5	77.8	71.9	65.5	70.6
1978–79	74.6	79.8	75.9	68.1	73.3
1979–80	76.5	81.7	78.7	70.9	75.3
1985–86	88.7	92.2	93.5	84.0	87.8
1986–87	89.6	93.7	94.0	84.5	88.4
1987–88	90.3	93.8	94.2	81.7	87.3
1988–89	98.8	91.1	91.4	80.0	85.2
1996–97	96.8	97.6	96.6	90.9	94.2

Source: Censis/Cnel (1980, 1997).

During the last few decades, the gap between the south and the centre-north, which was substantial in the 1960s, has decreased rapidly. *Scuole per l'infanzia*, therefore, though not compulsory, are almost universal both in availability and in attendance. The Italian *scuole per l'infanzia* are a universal service also because they are very low-cost for families who pay only a very small enrolment fee plus meals (Saraceno, 1998). The wide availability, low cost, and high attendance rate of children in kindergartens reflect the assumptions and ideas behind this service: all Italian children, starting from the age of 3, are expected to attend kindergarten as part of the normal educational curriculum. During the democratic period, the *scuola per l'infanzia* has been considered less a family policy than an educational policy aimed at reducing social class inequalities between children.

The day-care law of 1971 created high expectations, as it aimed to cover 6 per cent of all children under the age of 3 within five years. As late as 1991, however, this goal had still not been achieved at the national level. The day-care law established new state-level financial support to nursery centres (previously run by ONMI and by employers) organized and managed by regional and local governments.

Since 1971 public day-care services for children under 3 have been organized by local authorities (mainly by the municipality) and cater exclusively for children of officially working mothers, lone parents, and children under stress. Unlike kindergarten services, day-care services are not universal and are a 'service to be provided upon individual demand' and therefore optional, so that the user has to pay for at least part of it. Parents' payments towards publicly funded day care depend on their family income. Central government guidelines state that parents ought to pay, on average, 36 per cent of the cost. Parents pay less in certain areas, where local governments contribute more. Day-care services are open about 11 hours a day, five days a week, 11 months a year, and children typically attend for an average of nine to ten hours a day. As can be seen in Table 7, in 1992 there were only about six places for every 100 children under 3.

Table 7
DAY-CARE SERVICES BY GEOGRAPHICAL AREA, 1976–92

	North	Centre	South Isles	Italy
1976	3.2	2.6	1.0	2.1
1979	5.1	3.4	1.1	3.1
1983	8.5	6.5	1.5	5.0
1986	9.0	6.9	1.8	5.2
1989	9.6	7.5	1.9	5.6
1990	9.5	7.3	2.0	5.6
1992	10.1	7.5	2.1	6.0

Source: Based on figures supplied by ISTAT, various years.

The 1971 law on day-care services was more consistent with past traditions than was the law on kindergartens. Up to the present, little importance has been attached to day-care services. This is evident in the administrative carelessness in monitoring and providing data on the availability of and attendance rates in day-care services. Despite data limitations, great regional and municipal differences in public day-care provision are still clear. The reasons for the different expansion rates between kindergartens and day-care services for children under 3 in Italy are found in the different historical origins of these services. Indeed, the new democratic republic only partially changed past traditions and definitions of children's needs. In any case, it is important to point out that the law on working mothers, coupled with the day-care law, marks the first government effort to connect the arrangement of social structures with the conditions of working women. Nevertheless, the traditional division of work between genders within the family has been only minimally challenged by this law. It largely retained the ideas that a baby primarily needs his or her mother's care, and that the family is the best place to receive that care. The care of very young children (under the age of 3) remains a family matter, in which the role of woman as mother comes first.

Finally, in the framework of innovations in family policy and services of the latest governments, it is worth mentioning the National Plan introduced by Law 285/97.[9] This law introduces a National Fund for favouring the rights, the development, and the quality of life of children. Special attention has to be given to the improvement of the relations between children and parents, also to the improvement of parental and child care services, and to the institution of new after-school and leisure-time facilities targeted at children and parents. Funds are provided to local authorities (i.e. to regions and to the bigger municipalities) for intervention in favour of minor children, among which is intervention against family poverty.

The Spanish Case

An interpretation of the changes and continuity in policies related to the family and family responsibilities in the post-Franco period would be inadequate without considering the main institutional characteristics inherited by Franco's 'despotic corporatist' welfare state regime. First of all, in Spain insurance schemes were (and continue to be) related to employment. Therefore, social insurance benefits and services were mainly available to workers and their dependants, rather than to citizens in general. Furthermore, the welfare state granted (and still does) social security benefits according to different 'occupational status', while leaving social services almost completely underdeveloped.

The Survivors' Pension

Spain introduced widows' and orphans' pensions for the first time in 1955. The current survivors' insurance scheme was introduced in Spain in 1967[10] and since then it has remained almost unchanged.[11] The Spanish survivors' pension scheme is called 'dead and surviving' (*muerte y supervivencia*) and it provides different cash benefits to survivors of the deceased: funeral allowances, widows' pensions, orphans' pensions and provisions for other relatives.

Widows' pension (and since 1983 widowers' pension) This is paid to the widow and widower under certain conditions: the survivor is the current spouse, or, in the case of separation, the surviving spouse has rights to the pension independently of the cause of separation and/or divorce, according to the number of years they were married. Until 1983 widows' pensions were granted only to women aged 40 years or over; widows under the age of 40 only received the pension if they were unable to work and/or if they had dependent children entitled to orphans' pensions. As in Italy (until 1975), widowers received pensions only if they were unable to work and had been financially dependent upon their deceased insured spouse. Surviving spouses were entitled to 45 per cent of the actual or hypothetical retirement pension of the insured person. For widows under the age of 40 and widowers under the age of 65, entitlement to the pension was revoked if the survivor remarried or (before 1983) was no longer incapacitated and unable to work. After the Constitutional Court decision of 1983 on equality between widows and widowers, survivors could collect survivors' pensions regardless of their age, position in the labour market, or entitlement to other pensions.

Orphans' pensions All orphans under the age of 18 and all disabled orphans were entitled to orphans' pensions. Stepchildren were also entitled to orphans' pensions, provided that their natural parent had been married to the deceased for at least two years; there was no other family member legally obliged to maintain them (*obligación de alimentos*); they were not entitled to any other pension; and they had been living with and financially dependent upon the deceased before his/her death. Orphans are each entitled to 20 per cent of the pension of the insured, but the amount must not be below a yearly fixed amount.

Provisions for other relatives According to the Spanish social security law (since 1967), another group of deceased workers' family members might be entitled to survivors' pensions: orphaned siblings and nephews, mothers and grandmothers, fathers and grandfathers. In contrast to widows, widowers and orphans, however, their entitlement is connected to their level of need and to a number of other conditions. Relatives of the deceased claiming pension benefits had to have been cohabiting with and financially dependent on the deceased; without other relatives legally obliged to maintain them; and without any other source of income or pension. *All relatives* meeting the required conditions are entitled to 20 per cent of the deceased's pension.

Family Benefits: Long Immovable Family Allowance

The new democratic government inherited family benefits legislation from the Franco regime, and up to 1990 no relevant reform of family benefits took place. During the transition period and throughout the 1980s, family allowances were not reformed and, most important, not indexed to the cost of living. After 1966, family benefits were granted to all insured workers for dependent children and/or dependent spouses, but no longer for other dependent family members. Family allowances consisted – up to the 1990 reform – of both monthly and once-off allowances. From 1971 to 1990 families received monthly child allowances which amounted, for the general social security system (*Régimen General*), to 250 PTAs for each child under the age of 18. From 1971 to 1985 families also received a monthly spousal allowance for a dependent wife (or disabled husband) of 375 PTAs. One-off marriage grants of 6,000 PTAs were granted from 1971 to 1985, and one-off birth allowances of 3,000 PTAs were granted from 1971 to 1985 (López López, 1996).

Table 8
MONTHLY CHILD ALLOWANCES, SPAIN, 1970–92

	Minimum inter-professional wage (SMI)	**Allowance per child under age 18**	**Value (as % of SMI)**
1970	3,600	200	5.6
1975	8,400	250	3.0
1980	22,770	250	0.5
1985	40,140	250	0.5
1990	50,010	250	0.5
1991	53,250	3,000	5.6
1992	56,280	3,000	5.3

Source: Calculations based on data in González-Sancho López (1982) and López López (1996).

As can be seen from Table 8, the amount of the monthly child allowance was not indexed and did not change from 1971 to 1991; as a consequence, it has become negligible with time. In 1971, 250 PTAs constituted around 5.6 per cent of the *Salario Mínimo Interprofesional* (SMI, Minimum Interprofessional Wage), in 1980 the same amount constituted only 0.5 per cent of the SMI. At the aggregate level, this development was accompanied by a steep decline in expenditure on family allowances as a percentage of GDP and total social expenditure. From 1976 until 1990 real expenditure at constant prices fell to one-tenth (Meil, 1994).

The trend in family allowance expenditure is very similar to the trend evident in the Italian case. In 1970, in Italy social expenditure for family benefits was 14.1 per cent of total social expenditure; in 1980 it dropped to 5.8 per cent and by 1985 was only 4.8 per cent.[12]

Real family allowance reform in Spain did not happen until 1990, that is, 15 years after Franco's death. This reform introduced non-contributory social provisions in the Spanish welfare state. The reform introduced 'universal' child allowances for low-income families. This provision, however, was not related to the occupational status of the worker, but it was granted to all the population. In other words, Spain introduced, in order to support families with children, a means-tested provision similar to what Italy had introduced with its 'household allowance' in 1988. However, while in Italy 'household allowances' continued to be granted only to people included in the social security system, any low-income families in Spain could claim the benefit. In order to be eligible for the child allowance, parents must not exceed an income ceiling, which was raised by 15 per cent for each child after the second.

According to the law of 1990, child allowances in Spain are granted for dependent children under the age of 18 who are living with and economically dependent upon their parents (*hijo a cargo).* Children who are not considered dependent are those who, although under 18, are working or benefiting from a public pension, excluding the orphan's pension. Children who are at least 65 per cent disabled (*minusvalidos*) are also entitled to an allowance.

As a consequence of the 1990 reform, the value of the monthly child allowance greatly increased. In 1985 a child allowance was estimated at around 0.5 per cent of the SMI, while in 1991 it was estimated at around 5.6 per cent of the same wage. Obviously, the reform reduced the number of eligible recipients and excluded almost all dual-earner families. In addition, a recent study has shown that given that the amount of family benefit is not automatically adjusted for inflation, the value in real terms has fallen by 30 per cent since the last adjustment in 1991 (Bertelsmann Foundation, 2000b).

Spain only raised the amount of means-tested child allowances in January 2000. At this point the amount of child benefit was fixed at 48,420 PTAs (285 E) per year, which meant an increase of nearly 35 per cent. The same reform in January 2000 also introduced two new one-off benefits for large families only (a means-tested benefit for the third child onwards) and a non-means-tested multiple births allowance. However, this latter allowance seems to be only regulated in the contributory modality, therefore with similar contributory requirements to the maternity leave allowance and with limited coverage (Bertelsmann Foundation, 2000b).

Tax Relief for Dependent Family Members

The persistence of Francoist family allowance legislation in democratic Spain – from 1966 to 1990 – has been counterbalanced (but only in part) by deep changes in the tax relief for family dependants, which accompanied the introduction of the Personal Income Tax in 1978.[13]

Joint taxation became mandatory for married couples from 1978. This choice of the family as the unit of taxation was taken in Spain without much parliamentary discussion over possible secondary effects on women's employment. In addition, tax relief was introduced to take into account family dependants.

Tax credits (*deducciones de la cuota*) were granted in fixed amounts for marriage, which would be increased according to the number of family income earners by a fixed coefficient. In 1988 tax credits for marriage were abolished and replaced by a tax credit for joint taxation. Tax credits were granted also for children under the age of 25 (until 1989), and for other relatives (*ascendientes*) over the age of 70 who lived with the taxpayer or had a low income. Finally, tax credits were granted for disabled family members.

The fact that Spanish fiscal policy introduced tax relief not only for the wife and children, which is the case in most European countries (cf. below, appendix 3), but also for elderly and low-income parents is particularly interesting. This provision explicitly supports the family/kinship solidarity model. Spanish fiscal policies have emphasized the importance of tax credits for the elderly (over 70 years) and low-income earners *asciendientes* over time and have almost levelled the value between the two different kinds of allowances. Throughout the 1980s and up to the 1990s the value of tax credits for dependent elderly and dependent children has been almost the same (Salas and Perez Villacosta, 1992; González-Sancho López, 1982).

Table 9 shows that fiscal relief for the cost of children (now up to 30 years) lost much of their purchasing value during the 1980s. This and the latest legislative changes in favour of tax credits for *ascendientes* demonstrate that tax credits for dependent children have only partially compensated for the loss of public cash transfers to families with children.

Thus, during the democratic period in Spain the male breadwinner family model was not highly supported through family allowances or through tax relief. According to Meil (1994), the tax reform of 1978 in Spain was a mere fiscal reform and never became an instrument of family policy. From this perspective, the decision to treat the family as the unit of assessment, which was to the advantage of married couples, combined with the scant protection offered to the family through child tax credits, has to be interpreted as unintended consequences of a mere fiscal policy. In 1989 a constitutional decision[14] declared mandatory joint taxation for married couples unconstitutional. This decision opened political and social debates on fiscal policies and family models (Millet, 1992). Since 1991 the unit of taxation has been the individual. However, married couples (and only married couples) have the option of choosing joint taxation. Thus, in the field of the institutional acknowledgement of different family forms and different family models, the Spanish tax system has changed: married couples, *de facto* families and single-parent families are equally recognized (Millet, 1992).

Table 9
EVOLUTION OF TAX CREDITS FOR CHILDREN, SPAIN, 1982–91

	Tax credit 1 child (current prices)	**Tax credit 1 child (at 1985 price)**	**Index number 1985 = 100**
1982	12,000	16,348	108
1983	13,000*	15,738	104
1984	14,000**	15,235	101
1985	15,000	15,000	100
1986	16,000	14,705	98
1987	16,800	14,672	97
1988	17,600	14,666	97
1989	18,100	14,118	94
1990	19,000	13,888	92
1991	20,000	13,802	92

Source: for 1982 Sancho López (1982), for the period 1983–89) Salas and Perez (1992). Personal elaboration on the basis of consumer prices 1985.

* After the third child 18,000 PTAs
** After the third child 19,000.

Finally, the level of family benefits, both in terms of cash benefits and of tax benefits, has not only been diminishing over time, but is also very low in comparison with other countries (see appendices 5 and 6).

Maternity Leave Reform and Child Care Provisions

Maternity leave reform The law on maternity leave remained almost unchanged during the Spanish democratic regime, in strong continuity with the Spanish protective legislation which began at the turn of the twentieth century. According to the 1976 law,[15] women are entitled to a compulsory paid maternity leave of six weeks before childbirth and eight weeks after (paid at 75 per cent of her salary). In addition, mothers are entitled to an extended leave of up to three years (unpaid and optional), after the termination of the paid compulsory maternity leave. In 1989[16] the compulsory leave period was extended to 16 weeks (18 in case of a multiple birth), six weeks of which have to be taken after confinement. If both parents are in paid employment, the father is entitled to use the last four weeks of statutory maternity leave instead of the mother, as long as the latter's return to work does not entail any risk to her health. Since 1980[17] fathers may take the 'optional' unpaid parental leave instead of their employed wives. However, up to 1994 the law on 'optional' leave (now parental leave) did not guarantee that the job would be held for the mother (or the father). This unpaid parental leave is partly taken into account for old-age pension. The person who has cared for a child obtains one year's recognition towards his or her pension, per child (López López, 1994). In the same year (1994),[18] the eligibility requirements for maternity leave changed. Minimum contribution levels were reduced, and the overall benefits increased (maternity benefits passed from 75 per cent of mothers' normal pay to 100 per cent).

Also Spain, under the push of EU regulations, introduced new legislation in favour of maternity and parental leave in 1999. These measures allow for greater flexibility, better protection for expectant and breast-feeding mothers and better legal guarantees and new arrangements for parents in leave. Finally, existing unpaid parental leave and working time reduction entitlements are extended to take care of a relative in need of care (Bertelsmann Foundation, 2000a).

Child care policies During Franco's regime most day-care and pre-school services had been provided by the private sector, with the Catholic Church accounting for a large proportion of them. During that period, public child care policies had been extremely underdeveloped and mainly defined (after 1970) as pre-primary education: day-care services (*jardin de infancia*) for children aged 2 to 4, and kindergartens (*parvulario*) for children aged 4 and 5.

Fundamental changes in pre-school services began to occur in Spain only with the advent of democracy. After 1975 public pre-primary school services began to multiply, even though they mostly served the 4–6 age group. In 1977, as a result of the Moncloa Pacts, the leaders of the main political parties underwrote an agreement on education which aimed to increase the number of places in pre-primary school for children aged 4 and 5. In this period there was real political will to expand the number of state-run schools for children under 6 (Berea, 1992). As a consequence of these efforts, total enrolment in pre-primary schools rose from only 891,914 in 1970/1 to a record of 1,197,897 in 1981/2 (Palacios, 1989).

Two main features have characterized child care policies in Spain since 1975. First of all, the majority of child care policies are educational policies for children aged 3 or over and are run by the Ministry of Education and Science (Ministerio de Educación y Ciencia, MEC). Second, private child care or pre-school centres are *de facto* not regulated and controlled by the state; private centres serve most children under the age of 3. Before presenting information on the evolution of pre-school services (Table 11), a few remarks should be made about the data. First, there are no data for day-care services for children under the age of 1, although the compulsory maternity leave stops when the baby is 4 months old (Ministerio de Asuntos Sociales, MAS, 1992). Second, for children under 3, child care provisions (public and private) are underdeveloped. The absolute predominance of pri-

Table 10
CHILDREN ENROLLED IN PUBLIC AND PRIVATE DAY-CARE INSTITUTIONS, 1986–87

Age	Coverage	Public centres	Private centres		
			Total	Church	Other centres
0–1	na	na		na	
2	4.5	7.9	92.1	22.9	69.2
3	17.8	20.9	79.1	28.3	50.8
4	90.6	66.9	33.1	13.3	19.9
5	100.0	62.9	37.1	14.6	22.5

Source: MAS (1992). Calculations on the basis of MEC's data years: 1987–88. na = not available.

vate institutions (still mostly run by the Catholic Church) keeps this field of services outside the grasp of state regulation. Third, in the last 15 years the number of day-care schools and pre-school services run by private institutions has been drastically reduced, while the number of public day-care institutions and pre-school services has increased.

As can be seen from table 10, while children aged 2 and 3 are mostly cared for in private institutions, the situation changes for children aged 4 and 5. This age group enjoyed a very good coverage rate in public facilities in 1986–87.

Table 11
PRE-SCHOOL ATTENDANCE BY AGE, SPAIN, 1970/71–1992/93

School year	1970–71	1975–76	1979–80	1985–86	1990–91	1992–93
Under 1	na	na	na	na	na	0.6
1 year	na	na	na	na	na	2.8
2 years	3.0	5.6	5.2	4.7	6.9	8.3
3 years	11.5	15.0	15.3	16.7	27.8	44.8
4 years	43.0	51.4	66.6	85.6	94.1	96.9
5 years	69.0	81.9	87.4	100.0	100.0	100.0

Sources: Torres (1993) for 1970–71; INE (1977, 1982) for 1975–76 and 1979–80; MEC (Ministerio de Educación y Ciencia, 1988, 1994) for 1985–86 and 1990–91; Eurydice (1996) for 1992–93; Valiente (1996b), table 13.1 for under 1 and for 1 year. na = not available.

Religious institutions, private enterprises and co-operatives own and/or run private child care facilities. These display a high heterogeneity in both the prices and the types of services offered. Public institutions are run by the Ministry of Education and Science (MEC), municipalities, autonomous regions, and by the Ministry of Employment, which provides day-care services for children of working mothers.

Table 11 shows that the real increase in the number of children enrolled in pre-schools began in 1975/6, when most public pre-primary schools were set up. In 1985/6 all children aged 5 were enrolled in pre-primary school services, and since 1990/1 about 94 per cent of children aged 4 attended pre-primary schools.

In the late 1980s MEC opened a public debate on the education system, including pre-primary education. This debate resulted in a new education law, which passed in 1990 (organic law, no. 1, 3 October, Ley Orgánica de Ordenación General del Sistema Educativo, LOGSE). The 1990 law established that children aged 0 to 6 could participate in educational experiences (in contrast with the 1970 law, which made reference only to children age 2 to 6).

However, pre-primary schools continue to be defined as voluntary. The main problem of the Spanish pre-primary education system is the continuing existence of two well-differentiated policies. For children over the age of 3 the provision of places in public schools is quite high and mainly provided by MEC, within the framework of educational policy. A completely different situation applies for the under-3s. MEC has little involvement in public pre-primary schools dedicated to the under-3s: these are provided by regional or municipal governments or the Ministry of Social Affairs (MAS). MAS mainly provides for children from underprivileged families. Despite the principles inspiring LOGSE, the practices of MEC leave the control of the services for the under-3s to the private sector.

Several factors have hindered the development of public day-care services for children in Spain. First of all, most people have not challenged the 'traditional' division of responsibilities between men and women for the care of children. The entrenched belief that mothers are the best care providers for children under 3 has also been reflected in the low levels of female employment. Second, the private monopoly on day-care services for the under-3s has hampered any eventual public expansion. Furthermore, private institutions have also managed to divert regulation in this field. The Church is active in this area because it has always been interested in maintaining its control over education. Other private institutions oppose regulation because an imposition of higher quality standards for their services would likely force most day-care services to close down (Berea, 1992; Valiente, 1995). Analysis of the evolution of child care programmes and maternity law in Spain during the democratic period has shown that government policies have achieved important results in pre-primary school services for children aged 3 to 5. For very young children, however, there has been strong continuity with fascist-era policies and old ideology about women's roles. Most children are cared for within the family by women who do not participate in the labour market. What seems difficult to challenge are the ideas concerning children's needs and best welfare. The prevailing attitude is and has been that the child is better cared for within the family, by his or her mother. This is also supported by the optional *unpaid* parental leave that permits mothers (or fathers) to stay at home and care for children under the age of 3.

The Family Kinship Solidarity Model

This chapter has illustrated the main definitions of the family model that underlie certain social policies and social services. In particular, it has shown that during the democratic period in Italy and Spain the male breadwinner family model has been further stretched towards the family/kinship solidarity model.

Survivors' pension In terms of the definition of family models, the evolution of survivors' pensions in Italy during the democratic period – and especially the extension for the first time of the survivors' pension towards kin – produced two results. First, the introduction of kin entitlement in the 1950s and 1960s contributed to the explicit encouragement of the family/kinship solidarity family model. This extension also points to the assistance role played by survivors' pensions for elderly and low-income relatives, for whom the Italian welfare state provides no other form of protection. Second, the survivors' pension scheme supported the male/breadwinner role and single-earner family patterns until 1977.

In Spain, the evolution of the survivors' pension scheme shows two features that are relevant to definitions of the family model. First, the scheme upholds the previous purpose of protecting not only the wife and children of a deceased insured worker, but also other eligible dependent family members. This extension of the survivors' pension scheme towards other relatives points to the assistance role played by this scheme, as well as to the willingness of the democratic policy-makers to stick to the principles of the family/kinship solidarity model. Secondly, the Spanish survivors' pension scheme also supported the male breadwinner family model until sex discrimination was lifted in 1983.

Family allowances As far as the definition of the family model is concerned, the Italian family allowance scheme supported the male breadwinner family model until family law reform in 1975. However, family allowances lost their power to support the male wage as early as the 1960s. The latest family allowances reform seems to offer renewed support for certain institutional features of the 'male breadwinner' family model. In fact, the introduction of a means-tested measure tends to encourage the 'one-earner' family model among low-income families. However, the reform did not change the 'poor' level of support to families for the cost of raising children. Finally, despite the reforms of the 1980s, the fascist legacy of granting family allowances for relatives beyond a worker's nuclear family remains intact. This specific feature of the Italian family allowance system functions not only to sub-stitute for the lack of a national minimum income safety-net, but also to reinforce the family/kinship solidarity model by facilitating support to extended family members.

In Spain the family allowance system lost its importance as an instrument to support the male wage during the first decade of democracy. In other words, the state did not actively provide economic support for the male breadwinner until the reform of 1990. This law translated the policies into an assistance measure specifically targeted at low-income families and families with disabled children. Like the Italian case, the means-tested nature of current Spanish family allowances may foster 'single-earner' family patterns, which support the male breadwinner family model.

Taxation policies The other traditional instrument of male wage support, taxation policies, no longer favour the male breadwinner model based on single-earner families. In both Italy and Spain the unit of taxation has become the individual rather than the family. In Italy support for the definition of the male breadwinner model lasted until 1977, insofar as joint taxation was mandatory and the husband was named as the fiscal subject possessing rights and duties. After the individual became the unit of taxation, fiscal policies no longer supported the male breadwinner family model. This was particularly true, as tax credits for dependants, especially for children, were of low value. At the same time, the maintenance of tax credits for other dependent family members supports the argument that the Italian fiscal policy system incorporated certain assumptions of the family/kinship solidarity model.

In Spain the introduction of the 1978 tax reform partly supported the male breadwinner family model and partly supported the family/kinship solidarity model. The support for the first model can be observed by the fact that joint taxation was mandatory for married couples up until 1989. Mandatory joint taxation for couples tends to favour a male breadwinner family model. The support for the family/kinship solidarity model is evident in the fact that in Spanish fiscal policy, tax credits are granted not only for the wife and the children, but also for relatives of over 70 years of age, low-income elderly, and, since 1989, children under the age of 30.

Maternity laws and child care programmes In Italy and Spain child care programmes and maternity policies have not modified the idea that caring work is primarily a family (women's) responsibility. Especially in relation to public support for caring work, the analysis of child care policies in Italy and Spain highlights the fact that services for the care of children, especially for the under-3s, show continuity with past dictatorial regimes. In Italy, only 6 per cent of children between 0 and 2 were enrolled in public and private day-care services in 1992. In Spain in 1992/3, only about 0.6 per cent of all children under the age of 1 were enrolled in private and public day-care centres, while about 2.8 per cent of 1-year-olds were enrolled. There are no other public forms of care arrangements in these two countries; the Catholic Church has effectively opposed state and/or local government-run child care services.

Despite certain improvements in pre-school services in recent decades, and because Italian and Spanish pre-schools operate on a schedule which does little to accommodate working women in their efforts to combine family and paid employment, the family ultimately provides most of the child care for children over the age of 3 as well.

It can be argued that the 'stretched' male breadwinner for funnelling public resources to include kin is completed by a 'stretched' female caring model. Most young children in Italy and Spain are cared for within the family. Contributing factors to the in-home provision of care can include low levels of female labour force participation in the 'official' labour market; more flexible hours offered to women in the informal sector; insufficient day-care services. The most important overall factor is the availability within the family network of other women willing to care for children and/or for the elderly. This last factor is an important contributor to women's continuous career patterns in Italy and Spain. This does not mean that in these two countries all women having children can easily continue to work. There are also women who are obliged to leave their job to care for young children or sick family members because public provisions and services are inadequate and there are no other female relatives who can assist with the caring burdens. The lower rates of women's activity in the 'official' labour market, the idea that children are best cared for in the family, and the low development of public child care have mutually interacted to reinforce the expectation that women, namely mothers and grandmothers, will be the unpaid care-workers for families.

The future of the family/kinship solidarity model An important question is, will the most recent social policy and legal reforms change the Mediterranean family kinship solidarity model? Looking at more recent reforms, it is worth noting that means-tested benefits have increased in recent decades in all countries. The increasing of means-tested benefits is also a trend visible in the Italian and Spanish welfare states. As is well known, means-tested benefits tend to reinforce women's family dependencies, and perhaps the male breadwinner family model, because their level and/or right depends on the husband's income and discourages female employment (González *et al.*, 2000). However, although the use of means-tested benefits has also occurred in Italy and Spain, it is not clear that they will point to a strength of the male breadwinner family model. Rather more recently the focus on provision for low-income families with children and the explicit taxation reforms, within a context of youth and female unemployment and a high level of temporary and precarious jobs, will more likely reinforce the idea of long-term and extended family solidarity and obligations. While there is evidence of improving legislation enabling mothers to work (the recent introduction of new measures on parental leave), it is occurring in a context of a dramatic decline in fertility and reduced public resources.

NOTES

1. The male breadwinner model prescribes breadwinning for men and home-making for women. This model is supported by social policy and laws that encourage this gender division of labour through income benefits or indirect benefits for dependent family members.
2. Law no. 39, 18 January 1945.
3. The Constitutional Court by two declarations (17–26 June 1975, sentence no. 164; and 30 Nov.–6 Dec. 1979, sentence no. 140), declared unconstitutional that part of the law which discriminated between sons and daughters in relation to survivors' pension.
4. As far as the rights of women and men to survivors' pensions are concerned, the sex equality law of 1977 establishes in Articles 10 and 11 the equal position of the two spouses.
5. Tax credits (or tax deductions) refer to an amount which is subtracted from the taxes, while tax allowances refer to an amount which is deducted from gross earnings before the calculation of taxable income.
6. Law no. 825, 9 October 1971, and the following decree of 1973.
7. Article 11 of the law no. 860, 1951 states that the Labour Inspector may order the employer to provide a nursery near the workplace, instead of a nursing room, where the children can be fed and generally looked after. The employer may co-operate with other nearby factories or offices to set up a joint central nursery for all women workers of these different enterprises. This law originated during fascism (1934), when employers who engaged more than 50 women between 15 and 50 years of age were obliged to run nursing rooms or provide financially for the custody of workers' babies. On this issue see also Ballestrero (1979, 1996).
8. Law no. 53 of 8 March, 'Disposizioni per il sostegno della maternità e della paternità, per il diritto alla cura e alla formazione e per il coordinamento dei tempi della città' (Legislation for supporting motherhood and fatherhood, for the right to care and to formation and for the coordination of city time).
9. This law is entitled 'Measures for the promotion of rights and opportunity for childhood and adolescence' (Disposizioni per la promozione di diritti e di opportunità per l'infanzia e l'adolescenza).
10. The law is published in BOE (Boletín Oficial Español), 23 February 1967.
11. At present, survivors' pensions are elaborated in the general law on Social Security (Texto refundido de la Ley general de la Seguridad Social), 20 June 1994.
12. The indicators used for measuring the trend in expenditures for family benefits for the two

countries are not always the same. Given the lack of historical comparative data series for family allowances, I have chosen to rely on the existing published national level data instead of on the comparative tables published by the ILO and Eurostat. Comparative tables on social security contain very general headings that do not clearly describe what expenditures are included.

13. Law no. 44, 8 September 1978, 'Impuesto sobre la Renta de las Personas Fisicas' (IRPF).
14. Constitutional decision no. 45, 20 Februrary 1989.
15. Law no. 16, Article 25, 8 April 1976.
16. Law no. 3, Article 1, 3 March 1989.
17. Law no. 8, Article 46.3. 'Estatuto de los Trabajadores', 10 March 1980.
18. Law no. 42, Article 33, 30 December 1994.

Chapter 8

THE COST OF CHILDREN IN POLICY-MAKING PROCESSES

The final stage of this book involves selecting a crucial issue relating to social policy and studying it in depth. The selected issue is how the cost of having children and the collective responsibilities for sharing these costs have been acknowledged in political discourses and framed during the policy-making processes of four policies. The policies selected for the analysis are family allowances and child care policies in Italy, and tax deductions and family allowances in Spain. Family allowances, tax deductions and child care laws have been selected because in these areas the costs of having children and of sharing public responsibilities are 'usually' raised in political discourse.

Throughout the different historical periods and cultural contexts, the costs and responsibilities of maintaining children have been framed both in terms of public and private responsibilities, and within the family between mothers, fathers and/or other relatives. Family allowances and tax deductions are the two main policy instruments by which the state has acknowledged a part of the economic costs to fathers for the maintenance of children.

Child care law has been chosen because historically it originated as an acknowledgement of the cost of children in terms of forfeited wages and the time families spend providing child care for young children. Historically, the cost of care has been acknowledged in the form of public child care services, private care facilities, maternity and parental leave, and contributory benefits for the period of absence when caring for a child.

For each of these selected policies I will illustrate the main social and political actors involved in the policy-making processes, and the main discourses and ideologies they express. To this end the main legislative activities, the parliamentary debates and bills, and the legal texts (and to a limited extent collective agreements) of the four policies have been analysed.[1]

By looking at the policy-making processes of the four policies, this chapter intends to illustrate that in both countries the issue of the cost of children as such has never been fully tackled. Rather, the four policy studies will show that political and public discourses have only indirectly addressed this issue within debates on policies for different purposes. This helps to explain why in the democratic Italian and Spanish welfare states a low level of support for families with children is found.

The Case of Family Allowances in Italy[2]

In the sections that follow I will show how the cost of and the responsibilities for children have been conceptualized in family allowances. Family allowances are the main family-related cash transfer by which fathers have been compensated for dependent family members (children, wives and other relatives).

As mentioned in the previous chapter, in Italy fascist legislation on family allowances was not changed by the advent of democracy. The institutional features of family allowances remained unchanged up to the reforms of the 1980s, although in the long period before the reform of 1988 the policy issue was not absent from the agenda. On the contrary, from the first legislatures up to the reforms of the 1980s, family allowances had often been the object of parliamentary activities, resulting in a very fragmented system of laws and provisions. The core characteristics of family allowance legislation during the 1950s, 1960s and 1970s mirrored Italy's general approach to welfare state provision since the first legislatures: they were based on occupational status and developed mainly through the growth of fragmented provisions, without the framework of a comprehensive social policy system.

The purpose of this analysis of family allowances is to answer two sets of questions. The first set is related to policy-making. Who have been the main political and social actors in this policy? Why have family allowances, which aim to support the male breadwinner family model, gradually lost their importance compared to other social provisions? Finally, why did the Christian Democrats, with their many years in power and their explicit support for a 'family wage' system, fail to achieve family allowance reform? I attempt to respond to these questions through a review of the relevant legislative activities and a discussion of the attitudes of the main social and political actors involved in this provision.

The second set of questions concerns the conceptualization of the issue of the cost of and responsibilities for children. In what form, to what degree and when has the issue of the cost of children and of collective responsibility for sharing this cost become part of public political discourse? Who, in the mind of the law-makers, has been the best provider of economic support for children – the family, or the state? Within the family, which members have been considered responsible for providing maintenance for children? As far as this second set of questions is concerned, I focus my analysis on political discourse and ideology relevant to the cost of children as revealed in legislative bills, parliamentary discussions and collective agreements.

Allowances for Dependent Family Members: A Matter of Union Agreement

During the period of the first Italian legislatures (1948–63), family allowances were the object of legislative activities almost every year. However, the main impulse towards such massive legislative activity did not come from the legislative bodies, but from the bargaining agreements among trade unions, employers' organizations and the government. The 'centrist government' (led by Democrazia Cristiana, DC) introduced the reforms stipulated in these collective agreements into the parliament, and they often became laws without any legislative changes.

The parliamentary validation of bargaining outcomes on family allowances resulted in a lack of parliamentary debate over the issue of the cost and responsibilities for children. The collective agreements reconfirmed the idea that family allowances had to compensate not only for the cost of children, but more widely for the cost of dependent family members. However, it can be argued that parliamentary members shared a common point of view favouring the existing framework of family allowances, as well as the conviction that the issue, concerned mainly with workers' rights and labour policies, was best addressed through government, trade union and employers' collective agreements. In other words, during this post-war period family allowances were a matter of union and employers' agreement.

The Postponement of a Comprehensive Reform

The period after 1963 was marked by an important shift at the political level. The Socialist Party (PSI) entered into the government coalition, where it would remain until 1976. This political change had no effect on family allowance reform, however. Unlike the previous period, the importance of bargaining between the government and its social partners declined, thus family allowances were debated more frequently in parliament. Both majority and minority MPs underlined the need for an immediate reform of the family allowance system to extend coverage to more people and increase the value of the benefits. But debate on this topic took place only within the context of debates on more specific issues (such as the increase in the value of the allowance, the abolition of income ceilings, and the extension of family allowances to the agricultural sector). A general, open parliamentary discussion of extensive reform never occurred. At the end of the 1960s, under pressure from new social demands and partly as a result of the recent education law, a new issue entered the parliamentary debates: the education costs of children. This issue was raised several times, especially by left-wing parties, and stemmed from the desire to reduce educational inequalities between classes. But it did not produce any results in terms of family allowance reform.

Although political discourse since the 1960s reveals remarkable political consensus on the need for family allowance reform, there was also a wide gap between political discourse and political action. In particular, the DC avoided placing a comprehensive reform of family allowances on the political agenda and tried to gain opposition votes by introducing only minor changes and/or benefit extensions to other categories of workers. At the same time, the main opposition party, the Italian Communist Party (PCI), did not have the power to put comprehensive family allowance reform on the political agenda and did not block the minor changes initiated by the DC. This seems to highlight one of the most typical features of Italian policy- and law-making processes: approving only non-problematic questions and minor laws (*leggine*). A wholesale reform of family allowances was also unlikely during this period because law-makers were more concerned with issues leading up to the pension reform of 1968. In particular, trade unions (without opposition from the PCI) had agreed to postpone the push to abolish the contribution ceiling of family allowances in exchange for immediate pension reforms, thus precluding immediate family allowance reform. Thus reform was not the top priority for trade unions, nor for the main governmental party.

Supporting Low-Income Families or Supporting 'Wage Policy'?

As far as family allowance legislation is concerned, the 1970s continued to be marked by lively parliamentary activity, which was proposed both by the DC and by opposition parties. Most parliamentary bills introduced only partial increases in family allowances and only for specific categories of workers. However, in this decade for the first time the two main political forces, the DC and PCI, introduced in parliament two bills[3] whose main (and explicit) purpose was to reform the whole system of family allowances.

These bills provide some information about MPs' general ideas about family allowances, and indirectly also about the family model to be supported. The preambles of the two bills contain the traditional Catholic and communist sources of inspiration of social policy. The DC bill recalls the social doctrine of the 'family wage' expressed by Leo XIII in *Rerum novarum* and mentions the constitutional principles (arts. 31 and 36) that the Italian republic supports family formation, family functions and especially large families. The PCI proposal began, like that of the Christian Democrats, by referring to constitutional principles. In particular, this bill cited the paragraph from

Article 36 of the Italian Constitution which states that the wage of the worker has to be related to the quality of the job and 'in any case it must be sufficient to insure to the worker and to his family a dignified and free life'. But unlike the DC bill, that of the Communists emphasized the idea that family allowances were an instrument of 'wage policy' and were thus fundamentally related to trade unions' and workers' struggles.

The two bills provide different perspectives on family allowances. The DC proposals targeted the benefit to low-income families and focused on families with children, emphasizing the need to compensate families for education costs. The DC proposal also suggested gradually introducing a system of state-financed family allowances. By contrast, the PCI's perspective on family allowances was work-based. Their MPs regarded family allowances as a social insurance measure and a right of workers. The PCI recognized the high expenses of families with school-aged children, but also wanted to maintain the institutional features of family allowances, whereby benefits were paid to workers (the male breadwinner) for children, spouses and other dependent family members. Why, despite these comprehensive proposals by both the DC and the PCI, was family allowance reform not achieved during this legislature?

So far, the story illustrates not only the weakness of the political and social actors in favour of reform, but especially the ambiguous part played by the DC. Christian Democrat MPs said they were in favour of reforming family allowances and of economic support for families with children. But their behaviour indicated that they were pursuing other interests. For instance, they used the available family allowance funds to finance other social measures and delayed indefinitely a real reform of family allowances. At the same time, trade unions did not put family allowance reform on the political agenda either, since the unions had other issues at stake during that time (i.e. pension reform and day-care law).

In addition to both the weakness of the social and political actors and the interests at stake during this time, there are other elements which help explain the failure of parliamentary reform bills on family allowances in Italy. First of all, since the drafting of the Italian Constitution and up to the 1970s, all Italian political parties avoided reopening the discussion on 'family matters', because this issue highlighted the deep ideological division between Catholics and the Communist Party (cf. Chapter 5). Second, during the late 1960s and early 1970s, Italy was faced with important new social demands for expansion of the welfare state, for the introduction of provisions related to the care of children (i.e. day-care law in 1971), and for a reform of family law (the 1970 divorce law). In other words, there was no social demand for a reform of family allowances, and no social or political actors (family associations, trade unions) who gave priority to family allowance reform over other issues. Other factors also help explain the failure of parliamentary reform proposals. The Italian policy-making process featured a proliferation of laws targeting microsectors (*leggine*) that were often approved with the votes of opposition party deputies (Morisi, 1992). In this way it was possible to extend the family allowance system to agricultural and self-employed workers without undertaking wholesale reform and without openly discussing which 'family model' Catholics and communists wanted to support.

The First Step towards Reform: Initial Consensus on the Issue of Poverty

The legislative process of the initial family allowance reform began in 1983 with a decree law, which was then converted into law.[4] As the introductory text reveals, the 1983 reform stemmed from an agreement reached shortly before between the government, trade unions and employers' organizations. Consequently, family allowance reform was neither discussed in the framework of a family policy reform, nor inserted in a wider reform of social policy. Rather, it was framed within the context of urgent labour policies aimed at containing the cost of labour, and at favouring youth employment.

This initial family allowance reform, which eventually led to the broader reform of 1988, might be seen as continuous with Christian democrat reform proposals since the mid-1970s. Like the DC proposals, this reform focused on the cost of children for low-income families. In other aspects, however, it was quite distinct from the DC proposals, which favoured supporting *all* families, not only those of workers. At this time there was already a perceptible consensus on supporting low-income families, for instance, in the recommendations of the National Commission for the Study of Family Problems (MLPS, 1983) and from trade union declarations (CGIL, 1981).

This first reform, however, contained an internal contradiction stemming from the nature and legacy of the provision. The new supplementary allowances, though framed as a 'family wage' and conceived of as welfare assistance, were still granted only to dependent workers. This is hardly surprising, however, given that the bill was the result of a government–union agreement on labour policies.

Since the family allowance law was converted from a decree law and did not originate in parliament, the debates in the assembly were very limited and focused more on the instrument employed for achieving the reform (decree law, government–union agreement) than on its content. The cost of and responsibilities for children were not raised as issues during the parliamentary session.

The 1988 Reform: 'Household Allowance' as a Measure to Combat Poverty

Five years after the first lukewarm reform, the family allowance law was completely revised. Like the 1983 reform, the 1988 law introducing the reform began with the conversion of a government decree law.[5] After 40 years, family allowance reform was finally achieved through an 'emergency' legislative process. Moreover, the reform was inserted within a decree law containing a miscellany of different provisions. The 1988 reform abolished family allowances and established the 'household allowance', a completely means-tested measure. As in 1983 (but unexpectedly, given the rhetorical commitment of the DC to focus on the cost of families with children), the 'household allowance' was not targeted at families with children.

The parliamentary examination and discussion of the 'household allowance' took place in the committee before being sent to the full assembly. The assembly discussion was short and did not question the core of what the government had included in the decree law. The issue of the cost of children was raised, but not discussed as such. And once again open political confrontation and a parliamentary debate on the cost of children were avoided. Only some amendments concerning the need to compensate the higher costs borne by particular categories of families, such as those headed by lone mothers or with disabled children, were advanced in the Chamber of Deputies and inserted into the new law.

What have been the results of the reform of family allowances? The introduction of the 'household allowance' has resulted in the abolition of family allowances and in the introduction of a new measure aimed at protecting families against poverty. However, this new measure is a strange mix, neither capable of protecting all families with children, because it is a means-tested benefit, nor protecting all families against poverty, because independent workers and those not in the formal labour market are not eligible. The final reform and the abolition of family allowances resemble a typical 'Italian-style' compromise. In this case, the compromise was between trade union interests, a communist-oriented, work-based social policy model, and the Christian democrats' preferences for a social assistance measure.

The reform occurred 40 years after the end of fascism. It was a hidden reform in that it did not open public or political debates on the family nor the social policy model. Even well into the 1980s an open debate about the public and private responsibilities for caring work and for the costs of children was not possible. This study of family allowance policy illustrates that social and political actors' demands for a reform of family allowances were weak. Moreover, it shows that the issue of the cost of children was never raised successfully in the political arena, nor were the issues of family models or the division of caring responsibilities between the state and the family. In addition, the long law-making process of family allowance reform illustrates that other characteristics typical of the Italian legislative process were also important. The law-making process in Italy is slow, it is often concerned with microsectors, it brings together majorities that are rarely more than minimum winning coalitions, and it often proceeds with the vote of the opposition.

The Case of Law Instituting Publicly Financed *Nidi* in Italy

For many decades after the foundation of the Italian republic the responsibility and associated costs for the care of infant children in Italy remained with the family, and in particular, with the mother. This issue constitutes the domain in which the fascist legacy maintained its influence for the longest period. In the period following the Second World War, public law intervened in the matter of caring for young children only in the case of 'poor' families or towards certain types of families (i.e. lone mothers). The fascist-founded ONMI (Organizzazione Nazionale Maternità e Infanzia) welfare agencies provided for 'at-risk' babies and for the welfare of 'poor' mothers well into the 1970s. During the 1950s and 1960s the services continued to be considered a means of providing assistance to disadvantaged families (such as families with working mothers and families with disabled children). While families with a working mother pose an anomaly from the point of view of the ideal male breadwinner family model, it was a visible and widespread phenomenon, especially within the industrial working class.

The second focus of the analysis of the Italian case concerns the law on public day-care services for very young children (under the age of 3). The law passed in the Italian parliament in 1971 (law number 1044 on *asili nido,* commonly named *nidi*). It may be considered as the main policy case in which the state has at least partly recognized women's (mothers') work in care-giving. The analysis of the public day-care services (*nidi*) law focuses on the policy-making process at a specific point in time (1971). It studies how the caring work involved in raising a child, specifically an infant child (under 3), has been acknowledged not only as a family (mother's) or employer's responsibility, but as one to be shared and partly supported at the societal level. I will illustrate this case first by providing a picture of the policy-making process and then by focusing on the main discourses and attitudes concerning the cost of and responsibilities for very young children. The materials for the analysis are the bill proposals and statements of the main social and political actors involved in the policy-making process.

The Sociopolitical Context of the Law

During the late 1960s and the 1970s Italy underwent a period of social and political liveliness and a season of great reform. This period was marked by welfare state expansion, pension law reforms, and the important institutional reform introducing regional governments. In addition, the state passed a national law on kindergartens in 1968, as well as fundamental laws for women's emancipation and for the dismantling of the 'patriarchal' legal family model. It is worth remembering that after the constitutional compromise, the main political parties remained silent for twenty years on family issues.

It was in the 1960s that all kinds of groups began to demand action on issues concerning women's rights and families. The women's movement, women's sectors within trade unions, left-wing political organizations, social workers and education experts activated around the issues of women's rights and the need to assist with the costs mothers bear in caring for children. It was under the pressure of these new social movements, especially the women's movement and trade union demands, that the main Italian political parties were obliged to reopen the 'family matter'.

Public Day-care Services as a Woman's Right

This context of massive social demands and broad social reform produced the 1971 law number 1044 on *nidi*. This law was coupled with law number 1204 on working mothers that reformed the 1951 law on working mothers to include all dependent working mothers.

Debates on the *nidi* law began in parliament in 1968 and lasted for three years. However, the original proposal was initiated outside of parliament several years earlier. In 1965 trade unions and the Unione Donne Italiane (UDI, the women's organization, closely linked to the Communist Party) had both launched campaigns in support of a popular initiative bill (*iniziativa popolare*) that advocated child care for children under 3.[6] In the same year, under the pressure of social demands, PCI MPs introduced in parliament a bill for establishing a public system of *nidi*.[7] The two bills proposed to set up a ten-year plan for the institution of a publicly funded system of day-care services for the under-3s, and it favoured decentralizing the authority and the organization of such services from the central state towards local authorities (regions, municipalities and provinces, *regioni, comuni* and *province*). However, these two bills, which eventually were deposited in the senate, never reached the floor. According to Bimbi and Della Sala (1998), the failure of these bills was due to the recent formation of the 'centre-left' government coalition. In other words, the bills contained an issue that was too delicate, and the majority parties preferred to delay the issue until their coalition was more stable.

After three years of legislative inactivity, and under the pressure of the intense campaign launched by trade unions and the UDI, the process of policy-making on the *nidi* began again. At the end of 1968, during a period of political and social turmoil, three parliamentary bills were introduced in the Chamber of Deputies. The first was presented by the Communists, the second by the Christian Democrats, and the third by Socialists and Social Democrats, before their split.[8] However, all parliamentary members who introduced the bill were members of and acted on behalf of their corresponding trade union organizations: CGIL, CISL and UIL. The substance of the three proposals introduced in parliament was identical, thanks to a prior set of negotiations undertaken by the trade union organizations. Once submitted to the legislative body, the three bills were unified into a single proposal.

The bills proposed a five-year plan (1970–74) for introducing new municipal day-care centres (*nidi*) for the care of children under 3. In brief, the bills set up the amount and the forms of state funding for the *nidi*. A national committee was established for planning and controlling the funding and the allocation of tasks between the different public bodies involved, which was put under the control of the Ministry of Health. The bills also fixed the rules to be observed by the municipalities in setting up *nidi*, the criteria by which provinces would select municipalities to receive funding for the new service centres, and the role of the provinces in examining the municipalities' requests. In addition, the bills provided for the institution of mixed committees at municipal and regional levels. Finally, they established the share of the provinces, municipalities and employers in financing the system.

It is important to note that according to the bills introduced in parliament, the target *nidi* users were dependent women workers (*lavoratrici*), and their contribution to the cost of services was left to be defined by the municipality, according to the income of the users. However, given that there would at least initially be a gap between supply and demand, the bill established guidelines for the criteria to be employed in order to select workers. So, for instance, according to the bill, full-time working mothers and those who could not rely on the support of other family members would have top priority. The bills aimed for the creation of 3,800 new *nidi* in five years.

The growth of women's participation in the labour market, as well as the problems raised by increased geographical mobility, massive immigration, urbanization, and the scant availability of services for very young children were mentioned in all three bills as the main reasons for demanding the *nidi* law. What is striking in reading the introductions of the bills is the convergence of standpoints of the different political parties and the three different trade union organizations.

Not only was there complete convergence of positions on the main guidelines driving the law, but also on the priority to be given to the issue of women's employment with respect to other reasons justifying the creation of new day-care services. For instance, Christian Democrat deputies emphasized that while it may be true that a woman's wage was often needed by the family for economic reasons, the growth of women's participation should be considered an irreversible process driven not just by necessity, but by women's aspirations to participate in the professions. Perhaps a different weight was given to the issue of women's rights. PCI MPs stressed more than their colleagues from other parties that the bill aimed at improving working women's rights within the framework of union traditions. However, the introduction of the PCI bill stated also that once begun, the child care system had to be available for all women, independent of their work status.

Since the *nidi* bills were presented in the same period as attempts to reform maternity leave for working mothers, their purposes may be seen as a completion of the latter. In addition, it is worth pointing out that the bills were written mainly by female parliamentary members who belonged not only to trade unions, but also to different party-inspired women's organizations (UDI, close to the PCI, and Comitato Italiano Femminile, CIF, a women's organization close to the DC). Given the trade union backgrounds of the drafters, it is not surprising that the bills initially emphasized the difficulty working mothers had in meeting the caring needs of their children. The stated goal of these bills was to insure a woman's right to work.

The Shift from Women's Rights to Child Welfare

Once introduced in parliament, the three bills were merged and then remitted to a joint parliamentary committee on home affairs and public health (Commissione riunite: Interni, Igiene e Sanità Pubblica), which provided for the examination, the discussion and the final approval of the law.

The first discussion of the bill took place on 18 November 1971, two years after the presentation of the first bill in the parliament. The text submitted for discussion by the committee was the product of work done by a subcommittee. As far as the conceptualization of the care of children was concerned, the new text introduced important changes. The writers shifted attention from the issue of providing care for the children of working mothers to nurturing the children's psychophysical development. This new emphasis on child welfare seemed purposefully engineered by DC MPs to reaffirm in the discourse and the law that supporting child welfare meant supporting the primacy of the family and the mother's role in child care. The *nidi*, then, were not conceived of as the best places to provide for the welfare of infant children.

The two 'official' speakers designated by the committee were female DC MPs. The first, Amalia Miotti Carli, began her speech by arguing that the most important outcomes of the subcommittee's work had been the redefinition of the *nidi* in terms of child welfare. In the deputy's words, 'the assistance in the *nidi* provided for children under 3 constitutes a social service of "public interest", having as a primary goal the health and psychophysical development of the child. It is only in a subsidiary way that this service is aimed at facilitating women's access to work.' It can be noted that, as it was in the new text as well as in the rhetoric of the speech, the focus was placed on child welfare rather than on women's rights. But the analytical examination of the discourses of the two 'official' speakers shows something more. In the discourses of the DC MPs, the primary goal of the health and psychophysical development of childhood seemed first of all to indicate that *nidi* are a service choice of the last resort. In fact, Amalia Miotti Carli ended her speech by affirming that despite the presence of *nidi*, day-care services could never be conceived of as a substitute for the family and a mother's care. At-home child care should be the normal and most frequently used form of care for children from ages 0 to 3.[9]

The discourse of the second introductory speaker of the committee, Deputy Giannina Cattaneo Petrini, proceeded in the same way. These discourses, which remained unanswered by any alternative visions, made it possible to transform the original text in a way which reaffirmed the primacy of the family and the mother's role in child care.

Christian Democrat MPs raised other issues concerning the care of children. In particular, their discussion focused on the potential secondary effects of the public day-care services law. First, they suggested that the development of *nidi* might result in an 'institutionalization' (*istituzionalizzazione*) of children. Therefore, they advanced proposals which facilitated family participation in the *nidi*. Other DC members raised the issue that women's employment was too often the result of families' needs rather than of women's choices, and that the state should help provide alternatives to working outside the home for women with young children under the age of 3. No hostile or oppositional voices were raised during this discussion, which reaffirmed the central role of the family and of the mother in the care of very young children. It is particularly striking to note the total lack of intervention from the opposition, and the scarcity of contributions to the debate by PCI MPs.

The story of the parliamentary debates on *nidi* law raises two questions that are not easily answered. The first is why the joint committee had designated two Christian Democrat MPs as the 'official' speakers for the law? The second is why wasn't there any opposition to the shift of attention from women's rights to the primacy of the mother's role in children's care?

A consensual law-making process The committee met again some days after the initial presentation of the bill, and after a brief examination of the text the law was approved by the committee in the Chamber of Deputies for transmission to a committee within the senate, which approved the final bill in one session on 2 December 1971. The legislative process behind this law clearly shows *consociativismo*, a typical feature of the Italian legislative process whereby consensus among actors is reached before legislation reaches the complete assembly. Confining the legislative process to the autonomous work of committees and subcommittees limited the discussion and averted the risk of an open confrontation in the assembly. The closed law-making process permitted the change in the focus of attention from women's rights to child welfare (reaffirming that a child's welfare was guaranteed by a mother's care) without raising any discussion about the appropriate family model or more general social policy reforms. In the committee the law was examined and approved by the majority and the opposition together, without meeting any critics or opposition and without any further examination by parliament as a whole. *Consociativismo,* combined with left-oriented political parties' fears that they might not get sufficient votes to approve the law, explains why no opposition was raised concerning the deep change in the law's focus.

The Final Law: Who Should Care for Children?

According to the final law, 'the assistance in the *nidi* to children under 3, in the framework of a family policy, constitutes a social service of public interest' (law number 1044, Article 1). Therefore, in theory the final law conceived the *nidi* not as a welfare service targeted to children of working mothers, but rather as a 'universal' service fundamentally in the field of 'family policy'. This seems to place these services within a larger framework. But the second paragraph of the Article states: 'the *nidi* aims to provide for the *temporary custody* of children, in order to guarantee needs-based assistance (*assistenza*) to the family *and also* in order to facilitate women's employment in the framework of a broad social security system'. This second paragraph seems to emphasize the *temporary* and custodial nature of the service, that is, *nidi* is not defined as a service having educational goals, but rather as a measure of assistance towards families who cannot provide the care for their children. Then, but only as a secondary purpose, the *nidi* aims to make it easier for women to work. But, as argued by one Christian Democrat deputy, only in a subsidiary way. Thus, given the state's opinion that a child's welfare can be guaranteed only by the mother, emphasizing the temporary and custodial nature of the *nidi* over their educational purpose demonstrates the intention to assure what was considered the child's best welfare. This law, as well as the maternity leave reform law, was approved without general scrutiny of the underlying traditional family model. The roles of women and men in caring activities also remained unchanged. The PCI, then enjoying a period of great popularity, preferred to compromise with the Christian Democrats rather than to engage in an open conflict on 'family matters', especially those concerning the representation of women's roles.

It can be argued that in the final law the issue of who should care for children remained undefined, thus leaving unquestioned women's traditional burden of caring responsibilities. The law was not only ambiguous in defining its goals, but also in establishing what ought to be the public share of responsibility for the care of very young children. Finally, the very complex division of responsibility between the state, the regions and the local authorities in financing, organizing and running public day-care centres limited successful implementation of the law. The initial expectations and purpose of the *nidi* law – to cover 5 per cent of children in five years – was not achieved until twenty years after its introduction.

The implementation of the law on the *nidi* soon had to reckon with a strong internal contradiction: although it was defined as a 'universal' service, in practice parents had to prove their need in order to have access to the system. Because of the high demand for a low number of spaces, children of full-time working mothers, single mothers, divorced parents, and/or poor families had precedence. Moreover, because *nidi* was defined as a 'service upon individual request' rather than a universal privilege (like schools and kindergartens), state regulations required users to pay a portion of the cost of the service, calculated on the basis of the user's income. Even today the fees and criteria for means-testing vary widely between regions, and within regions, between municipalities. The availability of public nursery places varies from 30 per cent coverage in certain areas and/or regions such as Emilia-Romagna and Tuscany, to no coverage at all in Sicily. The national law on the *nidi* left substantial political discretion to local authorities. This has left room for very different degrees of implementation and for different representations of women's roles. In some areas this has resulted in a 'traditional' representation of responsibilities for child care, and in others, a new representation of the women's role and of the public share of responsibilities in caring for children.

The Case of Family Allowance Reform in Spain

The new democratic Spanish government inherited its family benefits system from the Franco regime, and up to 1990 no relevant reform of family benefits took place. As a consequence, the value of family allowances was eroded by inflation. Why did Spain not reform family allowances with the advent of democracy? Did family allowance reform and the cost of having children become an issue in the post-Franco period? Who were the main social and political actors?

The paragraphs that follow attempt to answer these questions, first by looking at the political, ideological and economic contexts of the reform, and then focusing on the process of Spanish family benefits reform. To follow the policy-making processes of family allowances in Spain is not an easy task, first, because the reform issue involved rearranging the entire Spanish system of social security. Second, it was only in the 1990s that the issue of family caring responsibilities became a matter of social policy debate independent of the social security system. Previously, it had been treated as labour policy, connected to 'worker status', and subject to the consent of trade unions and employer organizations. Third, family allowance reform discussions were always overshadowed by the economic crisis which affected welfare state reform, and by the need to cut the cost of social security. As a consequence, not only is it difficult to follow the policy-making process, but also the issue of the cost of children is somewhat scattered and not tackled as such.

The Transition to Democracy and the Burdensome Context of Reform

The initiation of democracy in Spain was coupled with economic crisis. After Franco's death all political parties needed to consolidate their electoral constituencies. The recession of 1973 and the social and political modernization of post-Francoist society forced heavy cuts in family benefits. During the transition period there had already been several social security reform proposals (i.e. the White Book of Social Security, Libro Blanco de la Seguridad Social, in 1977). All of them devoted a section to the social and economic protection of the family, and all pointed out the need to reform the family allowance system. However, none of these reform proposals had any real impact during the transition period.

The years immediately following the transition period – first led by the UCD government and then from 1982 by the PSOE – were marked by the continued formation of social pacts between the government, trade unions and employers' associations. After the Moncloa Pacts of 1977, the Acuerdo Nacional Económico (ANE) (Economic National Agreement) was created in 1981, under the UCD government. It established a tripartite commission to evaluate a plan for social reform, by which family allowances would also be reformed. However, the content of the pact and the commission's recommendations were subsequently neglected.

During the UCD government (1978–82) no great demand for family policy was put on the political agenda, and no changes were made in the field of social support for families with children, despite the UCD's platform declaring it to be in favour of economically protecting families with children. As was widely illustrated in the fifth chapter of this book, the main Spanish parties were divided between the centre-right, who strongly emphasized the institutional value of 'the family' and supported the idea of 'family unity' as well as policy interventions, and the left-wing parties, who avoided statements about the family and supported individual rights, legal changes to remove gender inequalities, and a feminist 'women's policy'.

During the socialist government (1982–93) economic problems remained a major concern to which the PSOE responded with neoliberal and neocorporatist policies (Guillén, 1996). According to Iglesias de Ussel (1994), political interest in the family issue and political will to act were lacking in the post-Franco period because there were other important problems that needed to be solved. In addition, the suggestion of family allowance reform brought back ghosts from the past. Several authors have argued that since Franco died there has been a trend in Spain to identify any sort of family policy with the pro-natalist, paternalist and patriarchal values expressed by the regime. The women's movement also added an important element of resistance to policies which seemed pro-natalist. The lack of consolidation of the democracy, the economic hardship, the weakness of social and political actors in demanding social policies for supporting families with children, the low legitimacy of family policy, and the ideological division on 'the family' (and on family policy intervention) between right-wing and left-wing parties are the contextual factors to be considered when looking at the policy-making process of family allowance reform in Spain.

The Initial Reform of 1985

The first family allowance reform occurred in 1985 in a context marked by the urgency of the pension reform. The social and political divisions and conflicts over pension reform were so deep that they resulted in a general strike in 1985 and in the break-up of the relationship between the governmental party, the PSOE, and the largest trade union association, the UGT (Union General Trabajadores). These struggles were followed by a period of further strikes and labour conflicts.

The initial family allowance reform was attained by law number 26, of 30 June 1985.[10] While both the bill introduced in parliament and the law in its final form intended to rationalize and reform the Spanish social security system, in practice it dealt mainly with pension reform. Despite the emphasis of most parliamentary groups on the need to reform the obsolete family allowance system, nothing happened. Throughout the writing of this law, the discussion of social and economic protection of families was very limited, and there was almost complete silence on the specific issue of the cost of children and how to share it.

Article 5 of law number 26, 'New Arrangements for Economic Benefits' (*reordenación de prestaciones económicas*) was devoted to family allowance reform. This Article resembled the proposals previously advanced by the government and, to some degree, by trade unions (González-Sancho López, 1982). The only difference is that the bill intended to increase family allowances for low-income families but within the social security system. During the drafting of the 1985 family allowance reform, only the opposition parties Grupo Popular and the Grupo Parlamentario de Cataluña expressed mild opposition to Article 5 and presented a few amendments to it. Although Socialist Party deputies professed to be in favour of the opposition parties' general idea to direct resources towards low-income families, and not only to insured workers, the majority rejected all of the amendments suggested to Article 5. Before looking at the second law, which completed the reform of family allowances in Spain, it is necessary to recall what happened between 1985 and 1990. Despite the 1985 law's provisions for changing the level of ordinary family allowances, the socialist government left the level of the benefit unchanged, resulting in a dramatic reduction of public resources directed at families.

The Road to the 1990 Reform: The Issue of 'Needy' Families

The second and final family allowance reform occurred in 1990, within a larger plan of poverty alleviation. It was accompanied by the introduction of non-contributory social assistance measures aimed at solving the huge problem of elderly people without any income. The following analysis focuses on the conceptualization of the cost of children during the passage of law number 26 of 20 December 1990 'establishing non-contributory benefits within social security' (*Por la que se establecen en la seguridad social prestaciones no contributivas*).

The government introduced the bill in parliament in December 1989, intending to found and administer a non-contributory form of economic benefit that would complement the 1985 family allowance reform. The PSOE, in power since 1982, had already expressed its intention to introduce a non-contributory system of pensions (PSOE, 1989). In particular, the Socialists hoped to extend certain economic benefits, such as old-age and invalidity pensions and child allowances, to citizens who had been previously excluded because of inadequate contributory histories. In their minds, this law would universalize entitlements to most economic benefits. However, the idea of universalizing family benefits was never framed in terms of benefits for *all* families with children, but always in terms of imposing selective eligibility criteria (mainly means-testing).

The part of the bill concerning the cost of children is found in Article 3, 'Changes in family benefits for children' (*Modificación de las prestaciones familiares por hijo a cargo*).[11] The bill acknowledged the monetary costs that low-income families bear in raising a child and the time workers spend caring for children when parents choose to take the unpaid 'optional' parental leave. The latter consisted of granting one year's contributions towards the worker's pension benefits, as well as protecting the worker's employment status during the one-year leave. This amended previous labour laws, which simply granted the right to an optional three-year leave.

As early as the presentation of the first amendments in the Chamber of Deputies, all the parliamentary parties (including the PSOE) agreed that the amount of ordinary child allowances fixed by the bill had to be improved. They also considered the benefits for disabled children to be insufficient. To fix these shortcomings, the legislators proposed amendments doubling the amount of ordinary child allowances and improving those for disabled children. The governing PSOE's new support for these expansions of the programme was probably related to the series of meetings they held with trade unions during the interval between the introduction of the bill and the presentation of the first amendments.

During the drafting of the law, only slightly different conceptions of child allowances emerged, while a more general debate on the issue of the cost of children and how the costs and responsibilities for children should be divided between families and the state never took place. As far as the family allowance policy is concerned, a large set of national and regional parliamentary groups, though sometimes employing different rationales, requested increasing the amount of each allowance and raising the income ceiling for eligibility.

The idea of increasing the number of beneficiaries of child allowances was strongly endorsed by the Grupo Popular. The deputies and senators of this parliamentary group affirmed that, according to universalization principles contained in the Constitution and pursued in the current fiscal system, it was unfair to restrict child allowances only to low-income families. The Grupo Popular MPs declared that the economic help for children ought to have a double function: to subsidize and to promote, and that the state's duty is to help families have the number of children they choose. Technically, the Grupo Popular's idea to universalize child benefits might have been achieved by drastically increasing the income ceiling for child allowances, thereby undermining the means-testing principles at the heart of the bill. The CiU (Convergencia i Unión), the Catalan centre-right party, called for changes similar to those requested by the Grupo Popular. They declared that the most desirable child allowance system would provide a fixed allowance for all children, regardless of family income. This party was the only one to frequently raise demographic issues.

In response to these challenges, the PSOE defended the fundamental principles of the bill, citing increased economic rationalization and its correspondence with the system of child tax deductions. For instance, during the conclusion of the drafting of the bill, Senator Ardanuy Costa said: 'I think we have to be economically realistic … this bill aimed to rationalize and to settle (*determinar*) … Therefore, we must set an income ceiling. It would be wonderful if all Spanish families received a substantial amount of child benefit, but this is not possible from several points of view.'[12] The final law incorporated almost all the amendments increasing the amount of ordinary child allowances for children under 18 and for disabled children over age 18.

The process of legislating family allowance reform in Spain shows that the demands of social and political actors for reforming this policy were weak. This explains only partially the delay of reform. The questions to be answered are: why were the demands for family allowances reform so modest, and why were they not translated into a wider debate over the cost of children for families?

For answers, we first looked at the context of the possible reform. In Spain the initiation of democracy was coupled with an economic crisis. In addition, it is worth reminding ourselves that political parties needed to consolidate their electoral constituencies after Franco's death. Political and social actors were also wary of 'family policy' that evoked any hint of Franco-era pro-natalism and anti-feminism. Another contextual factor important at the time of the 1985 reform was the end of the relationship between trade unions and the PSOE government. Trade unions, which in their declarations had shown concern for the issue of economically protecting workers' families, remained outside the decision-making process.

The need to reduce public expenditure and social security reform were the main factors in initiating debate and achieving legislation on family allowance reform. As declared by a PSOE MP, the policy had to be economically realistic. While the first family allowance reform aimed at rationalizing the social security system, the final law of 1990 attempted to keep the public's share of the cost of children as low as possible by targeting the benefits to low-income families.

The Case of Taxation Policy Reform in Spain

This section illustrates when and how the cost of children was taken into account within the framework of the drafting of Spanish tax reform, which in 1978 introduced the Personal Income Tax (Impuesto sobre la Renta de las Personas Físicas, IRPF). At the same time, I attempt to shed light on ideas about the issue of family responsibilities, especially towards children, which were endorsed by the deputies and senators. The analysis will illustrate the main social and political actors involved in the policy-making processes, as well as the main discourses and ideologies expressed through parliamentary debates, legal texts, and, to a limited extent, agreements between trade unions, employer organizations and the government.

The Concern for 'Family Unity' Versus the Cost of Children

After the introduction in parliament of the first bill concerning IRPF in 1978,[13] three issues regarding the family were raised. The first was the choice of the family as the unit of taxation, where the bill referred to the family as a 'unit' composed of the spouses and minor children. The second issue, related to the first, concerned the method of taxation applied to families in which there was more than one income earner. The third issue, which forms the focus of this analysis, concerned tax deductions (or credits) for marriage, for children and for other dependent family members.[14] The negative impact that the joint taxation system (mandatory for married couples) could have on women in the workforce was not much discussed, nor was the discrimination that this system introduced between married and unmarried working couples. The second issue was partly solved by applying a coefficient (1.3) to the number of income earners in the family, so that two income earner families were in part compensated. The third issue, the tax deductions granted for marriage and for dependent family members (children, elderly and disabled people) were the subject of some parliamentary discussions.

Tax credits for children and the concepts of family relationships that underlie them form the focus of the analysis that follows. The IRPF bill established, in Article 29 (*deducciones de la cuota*), a yearly tax credit of 5,000 PTAs per child per year (1.9 per cent of SMI), which was doubled for disabled children. Taxpayers could not claim tax credits for able-bodied children over the age of 25, for married children, or for children with an income over a given threshold, unless the child was one of the earners in the family unit. Article 29 also established the size of tax credits for marriage and for dependent elderly relatives. It is worth noting that during the drafting of the law, tax credits for children were not discussed separately from the other forms of family tax credits.

Two main issues were raised during the discussions about tax credits. The first concerned the amount of deductions for marriage and for the cost of dependent children and elderly relatives. The second concerned the conditions required for entitlement to the deductions. MPs' points of view on these issues are expressed in a number of amendments to the bill. These amendments are fruitful subjects of analysis, because in Spanish parliamentary procedure a proposed amendment is always delivered coupled with the reasons why a parliamentary group advances it.

Most of the amendments to Article 29 contained requests to increase the amount of tax credits and/or to expand eligibility for them. These amendments were introduced by a large range of political parties, from the PSOE to the Alianza Popular. The centrist party Unión de Centro Democrático (UCD), which held power, and the right-wing Alianza Popular (AP) proposed the greatest number and most far-reaching of the amendments. The deputies belonging to these two parties asked for larger tax credits for children and disabled children, as well as for a raise in the income ceiling above which the deductions were no longer granted. Both parties justified their amendments more with ideological defences of 'the family unit' based on marriage than with arguments advocating better support to families for the economic cost of children. Despite these similarities, there were also slight differences between the two political parties' proposals, especially regarding the reasons behind the amendments. The amendments sponsored by UCD MPs demanded that married couples not be put at a fiscal disadvantage to unmarried couples under the new tax law. According to UCD deputies, the number of unmarried couples was already increasing too fast. The party's amendments recalled Article 31.1 of the Spanish Constitution, which states that the duty of the state is to adequately protect 'the family' – that is, the one composed of two married spouses and their children. It is also important to note that the UCD was the only party to propose scaling the amount of tax credits for children according to the birth order of children. They advanced this proposal only three years after Franco's death.

Unlike the UCD, the AP wanted the unit of taxation to be the individual instead of the married couple. The argument behind this was the protection of families in which the wife was also a wage earner. In further amendments the AP also asked for increases in the amount of tax credits for children, for married couples and for the elderly, as well as a raise in the income ceiling for eligibility. The justification provided was similar to the UCD's: that 'the family' institution based on marriage had to be protected against increasing trends in households formed by unmarried couples. In addition, the AP hoped that increasing the deductions would minimize the effects of rising inflation on the purchasing value of tax credits. To sum up, while the UCD was mainly concerned with the protection of the family, the AP also emphasized the failure of the law to offer protection against the future erosion of the value of deductions due to inflation, and the potentially negative effect of the law on women's participation in the labour market.

Finally, the PSOE introduced only one amendment asking for a higher level (modest in comparison with the ones mentioned earlier) of tax credits for children. Like the UCD and the AP, they also asked to expand eligibility by raising the income ceiling. Their justification for this expansion rested on criteria of distributory justice.

To sum up, the drafting of IRPF tax credits for children were not conceptualized as measures explicitly intended to support the cost of children. At the core of legislators' concerns were neither the cost of children for families, nor the possible secondary effects the system could have on women's employment, but the protection of marriage and the 'family unit'.

In the final bill some of these amendments on tax credits for children (as well as for marriage and the elderly) were accepted. It is important to note that at this time the UCD was in power, and that most of the amendments had been presented by centre-right parties. The final law established a higher amount for the single tax credit for a child (6,000 PTAs, or 2.3 per cent of SMI) than did the first proposal (5,000 PTAs, or 1.9 per cent of SMI). In addition, parliament raised the income ceiling for eligibility.

Between Ideological Divisions: Inattention to the Education Costs of Children

Until 1989 the IRPF law (44/78) remained substantially unchanged, while, as mentioned in the previous chapter, the purchasing value of the single tax credit for children lost its value as a consequence of the high level of inflation (see above, table 10). The push towards tax reform stemmed from a constitutional decision, which declared that mandatory joint taxation for married couples was unconstitutional.[15] In the meantime, other social actors had expressed their demands for a change in the IRPF law. In particular, the main Spanish feminist institution of the central state, Instituto de la Mujer (IM), in its first programmatic document, Equality Plans (IM, 1987), requested reform of the gender-based discriminatory treatment resulting from the 1978 IRPF law.

Before starting an analysis of the long parliamentary process leading to the adoption of the IRPF law, I will focus on an earlier parliamentary action in which the issue of the cost of children emerged. The issue was raised by Grupo Popular the day before the introduction of the bill discussing the reform of the IRPF. It was raised through a bill of 'no-law' (*Proposiciones no de ley)* by which the government was asked to introduce into parliament a bill recognizing education costs through a sort of tax deduction.[16] The discussion was framed and rather obscured, however, by the ideological struggle between groups that supported the privatization of schools and the welfare state and groups that defended a public education system and welfare state. However, this bill was well-timed in relation to the discussions on changes in the IRPF law scheduled for the next day. It offered the opportunity to make further alterations in the section concerning tax deductions for children. The bill of no-law was rejected, but it garnered a notable level of political party support for compensating Spanish families for the cost of children. In particular, the proponents' arguments went as follows: there are two possible forms of fiscal support of the state towards citizens. On the one hand, the state can directly provide services (i.e. public schools), on the other hand, the state can allow tax credits for the cost of services not run by the state (i.e. private schools). What is interesting is the emphasis placed by a wide number of parties, both right-wing (i.e. Grupo de Minoría Catalana) and left-wing (i.e. Izquierda Unida), on the cost of education in general. Even the cost of public education (i.e. books, transport) constituted an uncompensated burden for Spanish families. According to the majority of the deputies who participated in the discussion, the Spanish fiscal system (as well as the IRPF) did not adequately take into account the cost of children for families.

The day after the rejection of the above-mentioned parliamentary intervention, parliament began the long process of making the IRPF law compatible with the Constitution's decision. The initial bill was introduced on 21 June 1989 and resulted in provisional law number 20 of 28 July 1989.[17]

This bill introduced some changes. First, a transitional period for tax declarations for 1988 and 1989, in which taxpayers had the option of filing either individually or jointly. Second, the bill abolished tax credits for marriage and formulated different tax brackets for both joint and individual taxation. The constitutional decision opened a general and political debate about fiscal policies and family models (Millet, 1992). When we analyse the drafting of the law, we note that the type of fiscal system and family models (especially, with reference to the effect of the fiscal system on married and unmarried families and on female labour force participation) was the object of parliamentary debates. However, tax credits for children during the drafting of IRPF law, despite the bill of no-law introduced by Grupo Popular, were not debated as a separate issue or even very much in relation to other issues. In other words, even during the IRPF reform, the cost of children did not become a prominent issue.

Further focus on the cost of children during the drafting of the bill was raised strongly only by Grupo Minoría Catalana, which introduced an alternative bill and a number of amendments. These aimed at protecting the family through the introduction of a 'splitting' system, which was more favourable to families. It also claimed the need to increase tax credits for children. The Catalan political party, within this context of reform, seemed to be the only party to have clear ideas about which family model they wanted, and the changes that current legislation needed at that time. It is worth mentioning that Catalonia is the only Spanish region to approve a comprehensive family policy law – in 1993 (*Pla Integral de Suport a les Families*, Generalitat de Catalunya, 1993).

An Acknowledgement of Youth's Long Family Dependency

In 1991 parliament revised law 20/89 in order to halt the transitory period of taxation. The parliamentary activity resulted in the final law 18/1991. However, as had been the case in the drafting of the provisional 1989 law, the issue of the cost of children was not debated as such except for some requests made especially by the Catalan party, Convergencia i Unió. This party asked to implement the constitutional principles according to which it is the state's duty to protect 'the family'.

According to the final law, tax credits are granted for each unmarried child (up to 30 years of age) who cohabits with the taxpayer, and who has no yearly income over the SMI. The amount of the deduction was fixed at 20,000 PTAs per year (2.8 per cent of SMI). This new amount did not exceed in value that of the previous decade. The most important change in the 1991 law was an increase in the age limit for tax credits for children up to 30, by which Spanish legislators have made an effort to adapt the law to the trend among young people to delay their departure from the family home. This legislative effort seems, however, not to have been present in the other important measure for supporting families for the cost of children: family allowances. In fact, in the latest family allowances reform of 1990, children's economic dependency on their parents is acknowledged only up to the age of 18. The new law which establishes that the amount of the tax credit for ascendants is doubled if the ascendant is aged over 75, also recognizes intergenerational ties and family support on the part of the Spanish welfare state.

A mild acknowledgement of the cost for the care of children can be seen, however, in the final 1991 taxation law. Here, parents are eligible for a tax deduction of 15 per cent of the child care expenses for children under the age of 3, up to a maximum of 25,000 PTAs per year, if both parents are working outside the home and their income is not higher than 2 million PTAs. This marks the first time the Spanish government recognized that parents might have costs for day-care for children, but only if both parents work. Otherwise, it was assumed that the mother would provide unpaid work caring for the children, which is assumed to be a no 'cost' service for the family. It is only when the care of a child is performed outside of the family – usually, given the low development of public services for the under-3s, by costly private day-care facilities – that a family can obtain a form of tax relief.

Comparing the Four Policy Studies

These four policy studies illustrate whether and how the cost and the responsibilities for children have been conceptualized during the policy-making process. Despite the differences among these policies, the processes which led up to them share an important commonality: the issue of the cost of children as such was never fully tackled. Rather, the political and public discourses have only indirectly addressed this issue within debates on policies for different purposes. Political groups and actors have introduced policies to protect the workers' wage or 'the family', reduce educational inequalities, contain the cost of labour, or combat poverty. The parliamentary debates and political discourses in all four cases show the incapacity and/or unwillingness of political actors to reckon with public discussions about the desired model of social policy, about family models, and about the definition of the relationship between the state and the family. As a consequence, it has been almost impossible to tackle the issue of the division of private and public responsibilities for the cost of and the care of children in public and political discourse. In the 1970s Italy had a good opportunity to readjust the division of responsibilities for the care and the cost of children between the state and the family, in a direction that would have relieved families, but it lost this chance.

The case of family allowance reform in Italy in the long term illustrates that social and political actors were not concerned with the issue of the cost of children. This helps to explain the delay of the reform. First of all they were concerned with the protection of workers and with the support of the family kinship solidarity model. The 'wage protection' of the worker was often an object of trade union concern in the 1950s and early 1960s. During the first decades of democracy, neither parliamentary debates nor collective agreements mentioned the cost of children for families separately from the issues of protecting the male breadwinner's wage and supporting family dependants. During the 1970s the issue of the education costs of children was raised in political debates, but only in terms of reducing education inequalities across social classes. These costs were neither faced nor acknowledged in family allowance reform. Next, in the 1980s the cost of children was raised again but within the context of 'rationalizing' the cost of labour and supporting households with low incomes. Finally, the traditional family allowances were abolished and transformed into anti-poverty measures for families. In other words, the final issue was poverty – not the cost of children or how these costs should be divided between the state and the family.

In the case of Spanish family allowances, there are both economic and political/ideological explanations for the delay of the reform and for the lack of open debate over the cost of and the public and private responsibilities for children. From an economic perspective, Spain's democratic transition was marked from the very beginning by deep economic crisis and the need to cut heavily into social policy expenditure. In this context, family allowance reform was seen as too costly and of lesser priority than other urgent social problems. This context of low legitimacy for policies towards families with children effectively silenced the social and political demands to increase family allowances and reform the system. In addition, there were deep ideological divisions between groups including the PSOE and the feminist movement and groups on the centre-left, on the one hand, and groups on the centre-right, on the other. The former suspected that any measure aimed at supporting families with children was really a disguised attempt to encourage and impose the traditional model of the family. The latter, indeed, intended to defend and to support the 'unity' of just such a traditional family model.

Comparing the current family allowance systems of Italy and Spain, it can be noted that Spain's final law introduced a mean-tested provision, as did Italy in 1988 with its 'household allowance'. However, while in Italy the 'household allowance' continues to be granted only to people included in the social security system, the Spanish benefit is extended to *all* low-income families. In addition, while Spain targets the benefit to families with children, Italy targets it to households with or without children. Finally, in both Italy and Spain the last reforms of 1988 and 1990 have increased the average value of the single allowance, but reduced the total number of beneficiaries.

The Spanish case of tax deductions for children illustrates that once a law is approved and the system has begun to function, further changes are difficult. The policy study on tax credits shows that while the system was promptly reformed because of the pressure of an external event (i.e. the constitutional decision), within the political arena it was difficult to widen the debate beyond the fiscal model of the couple. In particular, it was difficult to focus on the cost of children. Only one legal change has incorporated the issue of the cost of children for families: the tax deductions for dependent children up to the age of 30.

In the cases of family allowances in Italy and Spain and tax deductions in Spain, it can be noted that the main social and political actors could hardly get rid of the unsolved 'family matter' and overcome the ideological divisions about the family model they wanted. The latest reforms in family allowances show how quickly the costs of children fall by the wayside when trends toward 'rationality' and reductions in social expenditure are on the agenda.

The case of the law instituting publicly financed day-care services for children, the *nidi* law, is the only one in which there was a stated attempt to relate the issue of the cost of children to the wider acknowledgement of women's caring work. The analysis of this policy study shows that in this case the women's movement, trade unions and broader social demands were fundamental in putting this issue on the political agenda. However, this case is like the others in that there was no political will to enlarge the debate to the family model, to women's role in society, and to tackle the issue of the relationship between the state and the family in reproduction work. The focus of the parliamentary debate shifted, without any political opposition, from a new representation of women as both mothers and workers to the needs of children and women's traditional role in filling that need. The lack of political opposition to that shift can only be explained in light of the prevalent practices of '*consociativismo*' and by Italy's fragmented system of law-making. In fact, the law in its final form was designed without any discussion in the assembly, where differences over content are more difficult to resolve. The final law upholds the traditional representation of the family and of women as the best providers of care for the very young. Finally, the policy study on the *nidi* law shows the lingering ambiguity about who should pay and care for children. This lack of clarity at the national level has contributed to decades of uneven implementation at the local level.

NOTES

1. The references to all parliamentary documents used are, for reasons of space, not included in the text. For a full and complete reference to all parliamentary documents and sources used, see Naldini (1999).
2. Part of the material contained in this chapter, in particular that regarding family allowances in Italy and Spain, has been already published in Naldini (2000).
3. The first bill I have analysed was introduced in the Chamber of Deputies by Christian Democrat MPs, deputy Anselmi and colleagues, on 17 February 1971, bill no. 3086, 'New Legislation for Family Allowances' (Nuova disciplina degli assegni familiari). The second bill was introduced in the Chamber of Deputies by PCI Deputy Pochetti and colleagues on 19 May 1971, bill no. 3400, 'The Increase in the Amount of Family Allowances and the Modification of the Legislation of 30 May 1955' (Aumento degli assegni familiari e modifica delle norme del testo unico approvato con decreto del Presidente della Repubblica 30 maggio 1955, no. 797).
4. Law no. 79, 25 March 1983 entitled 'Law Converting the Decree-Law of 29 January 1983, no. 17, Concerning Measures for Containing the Cost of Work and for Encouraging Employment' (Conversione in legge del decreto-legge 29 gennaio, 1983, no. 17, recante misure per il contenimento del costo del lavoro e per favorire l'occupazione).
5. Law no. 153 of 13 May 1988, 'Law Converting the Decree-Law of 13 March 1988, no. 69 Concerning Provisions for Social Security and for Improving the Management of Harbour Bodies and other Urgent Provisions' (Conversione in legge, con modificazioni, del decreto-legge 13 marzo 1988, n. 69, recante norme in materia previdenziale, per il miglioramento delle gestioni degli enti portuali ed altre disposizioni urgenti).
6. The popular initiative bill was entitled 'Institution of Social Services for Children Under 3' (Istituzione del servizio sociale degli asili-nido per i bambini fino a tre anni) (Atti Parlamentari Senato, APS, IV Legislature, disegni e relazioni, no. 1043, 26 February 1965).
7. The first bill was presented by Senator Minella Molinari Angiola (PCI) *et al.* on 23 January 1965 and was entitled 'New Legislation for the Assistance of Maternity and of Primary

Infancy and Ten-Year Plan for the Development of Day-Care Services' (Nuove norme per l'assistenza alla maternità e alla prima infanzia e piano decennale per lo sviluppo degli asili-nido) (APS, IV Legislature, disegni e relazioni, no. 967, 23 January 1965).

8. Accordingly, the bills were the following: nos. 796, 805 and 982, all having the same title 'Ten-Year Plan for the Institution of Municipal Asilo-nido with the Financial Participation of the State' (Piano quinquennale per l'istituzione di asili-nido comunali con il concorso dello Stato), (Atti Parlamentari Camera, APC, V Legislature, disegni di legge e relazioni, 23 December 1968).
9. The speech is made by Deputy Amalia Miotti Carli (DC), the official speaker for the II Committee (APC, V Legislature, Commissioni Riunite, 2° Commissione (Interni) e 14° Commissione (Igiene e Sanità Pubblica) 18 November 1971, p. 28).
10. It was entitled 'Urgent Measures for the Rationalization of the Structure and for the Protective Action of Social Security' (Medidas urgentes para la racionalización de la estructura y de la acción protectora de la seguridad social).
11. Boletín Oficial de las Cortes Generales (BOCG); Congreso de los Diputados (CD), IV Legislature, Proyecto de ley, Serie A, no. 1-1, 22 December 1989.
12. BOCG, Senado (S), IV legislatura, Boletín no. 40, Diario de Sesiones, Pleno, 7 November 1990, p. 2095.
13. BOCG, CD, I Legislature, Boletín no. 46, Proyectos de ley, 11 January 1978.
14. As explained in the Spanish case, tax credits (or tax deductions) refer to the amount which is subtracted from the taxes.
15. The discrimination between two-earner and single-earner married couples had been partly faced by a 1985 law which introduced variable deductions for families with more than one earner.
16. The original title of the bill is 'Proposiciones no de ley del Grupo Parlamentario de Coalición Popular'. This asked the government to introduce into the parliament a change in law 44/78 of 8 September, 'del impuesto sobre la renta de las personas físicas' (IRPF), granting a tax deduction to families for the cost of education (BOGC, CD, IV Legislature, Boletín no. 200, 20 June 1989, p. 11344–51).
17. BOCG, CD, III Legislature, Proyecto de ley, Serie A, 122 I, 21 June 1989.

CONCLUSION

This book develops within the theoretical and empirical body of comparative welfare state research, to which it contributes both in terms of empirical information on specific historical characteristics of Mediterranean welfare states, and in terms of further developing the insights provided in this field by gender-oriented studies.

The hypotheses behind it are two-fold. First this book asserts that the balance of responsibilities between the state and the family in the area of reproduction and caring work is a crucial issue for understanding variations among welfare states – especially the particularities of the Mediterranean model. The assumption is that southern European welfare states allocate responsibilities between the state and the family in a way which marks the distinctive functioning of this model. The second hypothesis stems from a critical approach to the current theory of gender and the welfare state. It is grounded on the premise that variations across welfare states and family models cannot be fully understood if the analysis is confined to how state policies have affected gender relations and vice versa. Rather, these variations reveal that in certain countries not only gender relations, but also intergenerational and kin relations, are important elements for grasping the different ways welfare states develop and function.

The origins and the development of state policies towards families are not separate from the national development of welfare states. In fact, they are instruments aimed at solving the common problems that motivate the general expansion of modern welfare states in the first place (i.e. the process of industrialization and the related problems of modernization, the growth of social and political movements and the new demands coming from society). However, cross-national variations in state policies towards families also show that there have been issues (i.e. poverty, fertility decline) which, although common to most European countries, were addressed differently across the diverse social, economic and cultural contexts. It particularly concentrates on the fact that welfare states over time and across countries have not been neutral towards the family. Welfare provisions were introduced on the basis of a set of assumptions about the gendered division of roles within the family and in the labour market; assumptions about the 'ideal' role of women and of men in society; and assumptions about the functioning of the kin system. These assumptions in turn have contributed to strengthening specific family/kin systems.

The evolution of state policies towards family responsibilities in Italy and Spain from their origins to the end of the two authoritarian periods shows that policies in these two countries have differed from those of other countries, and that there were important timing variations between the two countries. In certain historical periods state policies focused on women's responsibilities, while in others they stressed men's responsibilities and/or the dependent status of the wife rather than the dependency of children and other relatives.

The first phase of state policies towards families in both Italy and Spain, as in most other European countries, originated at the beginning of the century. States first paid attention to women's family responsibilities through maternity leave and maternity benefit. Further expansion of Italian policies for family responsibilities developed a few decades later, between the two world wars, concurrent with changes in other European countries. This second phase was marked by a shift in state support from working mothers to male breadwinners, for their presumed responsibilities for dependent wives and children (during the 1920s in Italy, and the 1940s in Spain). The time-lag between Spain and Italy was strictly related to the later development of social security schemes in Spain. Women's and children's status of dependency on their husbands/fathers was embedded in the first social security schemes (e.g. old-age pension schemes). That is, the issue of familial dependence and the state position *vis-à-vis* dependent family relationships was first solved through the introduction of derived social rights (i.e. survivors' pension for widows and orphans). This approach mirrored those of many other European countries.

The twenty years of the Italian dictatorship and the first decades of the Franco regime further strengthened wives' and children's financial dependence on their husbands/fathers. But at that time, the status of the dependant was addressed less in terms of derived social rights, and more in terms of state income support to the male industrial worker – the male breadwinner. It is well known that government support to the male breadwinner for family maintenance was a common approach in the majority of industrialized countries at that time (i.e. Great Britain, France and Germany). A unique characteristic of the Italian and Spanish cases is that state policies 'stretched' public support to the male breadwinner beyond the dependent wife and children to include parents, siblings and other relatives. However, it must be kept in mind that this extended concept of family responsibilities corresponded to the persistent notion of legal kin obligations, which dated back to the nineteenth century in Italian and Spanish family law.

The experience of fascism is important because it marked a clear shift in state policies towards family responsibilities. For the first time in their histories, Italy and Spain introduced an all-encompassing system of family policy having strong pro-natalist aims. The legal definition of family relations and the traditional family law systems were not changed much apart from the strengthening of the abortion law, the marriage law and a further reinforcement of the definition of family relations. What really changed with the advent of the two dictatorships were the conceptualizations of the family as an institution and the development of family policies. The family was regarded as an institution organically linked to the state, of which it was both an instrument and a constitutive element. The regimes of Mussolini in Italy and of Franco in Spain supported a traditional (Catholic) and 'corporatist' family model. The man was regarded as the breadwinner, legally defined as the head of the family, and was helped in his duty as family provider through the introduction of a set of family policies which ranged from family allowances to tax benefits. The woman was assigned mainly to home-making, and discouraged from entering the official labour market. At the same time, women were also regarded as having an important social function in the areas of reproduction and of continuing traditional authoritarian values, both within the family (between genders and generations) and within society. In light of this important social reproductive function, one can understand why the two regimes, while maintaining the dependency of wives on their husbands, also developed certain welfare provisions for women and their children, and recognized the social rights of single mothers.

One of the main findings of this book is that in the two historical cases the male breadwinner family model was never wholly achieved. The analysis of different socio-economic characteristics of these two countries in comparison with other European countries explains in part why the male breadwinner family model was achieved only to a lesser degree. In fact, while state policies during the two dictatorships aimed at supporting the male *industrial* wage worker (as in most other European countries), they were applied to economies primarily having an *agricultural* structure. Therefore, there were few dependent industrial workers actually entitled to these benefits. In addition, the modernization processes of these two countries have been extremely uneven and have been accompanied by heavy internal and external migration, structural employment shortages, and the development of small enterprises within agricultural settings. While very few women were in the official labour market, most of them had to work for pay and held marginal jobs as agricultural workers or domestic helpers in the informal labour market.

These two countries also share some cultural factors which help explain the partial adoption of the male breadwinner family model. First of all, as mentioned above, both Italy and Spain were primarily rural societies until well into the twentieth century. In Italy, rural inhabitants and workers were still the majority in the 1936 special census, and in Spain the agricultural sector was predominant until the 1960s. In this context the norms related to the institution of marriage and family values were Church-oriented, and intergenerational obligations and kin solidarity were expected and perceived as having strong moral value. Finally, nineteenth-century legal definitions of family and kin obligations remained in effect in twentieth-century Italy and Spain. In sum, the socio-economic structure, the cultural traits of the Spanish and Italian family systems, and the persistent legal definition of family and kin obligations were factors that favoured less a male breadwinner family model and more a family/kinship solidarity model.

State policies aimed at supporting the male breadwinner that were introduced in the early Italian and Spanish social security schemes took into account not only the nuclear, urbanized and bourgeois model of the family, but also the 'rural' and traditional family model based on intergenerational ties and kin solidarity. These conceptions of the family are visible in the *assegni familiari* in Italy, from which the male breadwinner received benefits for parents, siblings and parents-in-law with no or low income. They are also expressed in Spain's *subsidio familiar*, *plus de carga familial* and the health insurance system, where benefits are extended to dependent relatives beyond the nuclear family. However, although their reference model was that of the agricultural family, it was through the figure of the male industrial worker that state policies towards family responsibilities were extended to other kin members, and that social protection benefits were funnelled to a greater number of people.

In conclusion, during the two dictatorial periods, Italian and Spanish laws and social policies have encouraged a male breadwinner family model that is 'stretched' towards a family/kinship solidarity model. State policies for families in Italy and Spain assumed that the breadwinner was responsible not only for a wife and children (that is the nuclear family), but also for other dependent family members.

During the democratic period the Italian and Spanish welfare states maintained most of the past characteristics belonging to the male breadwinner family model extended towards a family/kinship solidarity model. Italian and Spanish civil laws maintained the traditional strong legal definitions of family and kin obligations, while social and fiscal policies continued to support certain features belonging to the family/kinship solidarity model. They have not modified the mechanism of funnelling a part of public resources through the male breadwinner to dependent family and kin members. Survivors' pensions, family benefits and tax credits still reach dependent family members through wage workers. The perception of caring work as a family (woman's) matter to be met through intergenerational ties and kin solidarity has also remained unchanged in the democratic period.

The most relevant changes during the democratic periods are found in the weakening of supports to the male breadwinner at both legal and social policy levels. Thus, in Italy the reduction in the value of support to fathers for the cost of children (family allowances, tax deductions, and so forth) since the end of the 1960s has diminished resources to fathers without easing their economic and legal obligations. The abolition of the legal figure of the family head in the reform of the family code in 1975 took away any legal basis for a superior position of the father/husband in the family. The same occurred in Spain in the 1970s and 1980s. However, unlike in other countries, the Italian and Spanish male breadwinner model was not greatly weakened and/or replaced by changes in women's behaviour, by increased female labour force participation, or by increasing marital instability. Instead, Italian and Spanish husbands and fathers were still perceived as being the main breadwinners. While they were privileged accordingly in the labour market, they were left without some of the earlier policy supports that facilitated this role.

In Italy and Spain the most relevant change that has weakened the male breadwinner model was neither the rise of the dual-earner family model – which has gained importance in recent decades in these two countries – nor the removal of gender inequalities in social policies and legal norms. Instead, it has been the lower level of public financial support to fathers for their children. The minimal public support for families with children is notable in Italy and Spain, both relative to other European countries and compared to support in each country during the authoritarian periods. Thus, since the last decades the male breadwinner family model in Italy and Spain has lost one of its fundamental pillars: income support to the father's wage for the cost of children. Finally, this book shows that the male breadwinner family model can assume a different weight and diverse shapes across time and across countries, and, thus, that there are various definitions of family models which change over time.

Although the male breadwinner family model has been weakened in terms of the level of public support to fathers for children, it is still used as a means of distributing 'poor' resources and to compensate for imbalances in the distribution of social protection throughout the kin network and between genders and generations. This mechanism of redistributing resources by 'stretching' the male breadwinner family model has provided protection to most of the people who are not in the labour market and/or have not had a steady job. In this light, the current two-fold protection system can be regarded not as the consequence of the residual character of the Mediterranean welfare state, but as the result of an interplay between the occupational model of development of social policies, the specific characteristics of the labour market, the functioning of the family/kin system, and assumptions about the functioning of intergenerational ties and kin solidarity as a means to assure financial support and caring work to individuals.

Family changes in Italy and Spain show significant similarities: in the family formation phase, in the decline of fertility, in the lower level of family dissolution, and in the long permanence of the young in their parents' home. In other words, they show similar family patterns as well as similar trajectories of family changes. One of the most important similarities between the two countries, which differentiate them with respect to other countries, is that intergenerational ties and kin solidarity are still the hallmark of the functioning family system. In the Mediterranean welfare state a huge amount of reproduction and caring work is still guaranteed by the functioning of the family system. In addition, this welfare state model has developed largely on the basis of a set of assumptions in which intergenerational ties and kin solidarity are expected and supported. The continued relevance of intergenerational ties and kin solidarity for social welfare in these two countries helps us understand not only the continued reliance on the 'stretched' male breadwinner family model, but also the lack of certain social policies and/or the unequal distribution of rights, benefits and public provisions.

In addition, the 'stretched' male breadwinner is completed by a 'stretched female caring model'. In these two countries child care programmes and maternity policies have not modified the idea that caring work is primarily a woman's responsibility. The lower female activity rate in the official labour market; the presence of intergenerational and kin ties; the prevailing idea that children are better cared for in the family; and the low development of public child care provisions have helped reinforce the expectation that caring work should be provided by the unpaid work of a woman within the family and/or the kin network (i.e. grandmothers). Women who do manage to establish continuous working careers generally have other women within their kin network who are willing or able to assume the working woman's caring responsibilities.

The withdrawal of public support for families with children during the democratic period in Italy and Spain, the traditional strong definition of family obligations, and the continuing importance attached to dependent kin relationships in certain social policies targeted at the male breadwinner have contributed to reinforcing expectations and assumptions about intergenerational ties and kin solidarity along the life course, both within the household and between households. This analysis shows that the redrawing of the boundaries between state and family responsibilities has placed the bulk of work for reproduction and caring work on the family and even increased family responsibilities.

The scarce state support for families with children, and the absence of political and social demands for family policies, has paradoxically been relatively uncontested in a society characterized by high expectations about intergenerational ties and kin solidarity.

This book has attempted to explore the relationship between state policies for families with children and family solidarity. It concludes that the scant attention on the part of the state towards family responsibilities during the democratic period is not caused by the permanence of strong kin solidarity. Instead, the latter, along with the permanence of strong legal definitions of family and kin obligations, has resulted in greater expectations that families should meet their own needs.

The main explanatory factors that account for the family/kinship solidarity family model during the democratic period are not identical for the two historical cases, but they have certain similarities.

Occupational Structures and Labour Market Features

Italian and Spanish occupational structures have remained different from or have changed in different ways compared to other European countries, both before and following the demise of the two dictatorships. Certain specific traits of the southern countries' modernization processes have prevented the *de facto* achievement of the male breadwinner family model, and they have hampered the achievement of full male employment. These distinctive economic qualities have included large territorial imbalances, internal migration, structural employment shortages, the presence of a widespread informal economy, the development of small enterprises in agricultural settings, low female participation in the official labour market and a high number of precarious and 'black' jobs for both women and men.

The Legacy of Fascism

An explanatory variable relating to the low support of Italian and Spanish states for families with children has been found in the legacy of fascism. Despite the different political configurations and chronologies of authoritarian rule of each country, this legacy rendered it difficult and unattractive for any social or political force to formulate public policy supporting families for the cost of children in the new democracies. In Italy, the fascist experience left strong opposition towards any policy that seemed to have pro-natalist aims. Policies explicitly aimed at supporting families for the cost of children were easily perceived as pro-natalist, and were thus avoided. The same was true for Spain, where law-makers tried to avoid policy-making in the area of the family during the post-authoritarian period. Interestingly, the analysis of the two historical cases shows that the refusal of new policies suspected of demographic aims did not lead to the dismantling of previous policies (i.e. family allowances) that supported the cost of children. Instead, these policies were simply left to lose purchasing power and value over time.

Political, Ideological and Cultural Configurations

Italy and Spain display different political configurations. They were characterized by different political party systems, and by strong diversity in the timing of their democratic transitions. The DC led the Italian government in powerful coalitions for more than 40 years. In contrast, a national Christian Democrat Party never formed in Spain. Spanish politics after the transition period was marked by a long period of left-party power enjoyed by the PSOE. In Italy the main left-oriented party (the PCI) was excluded from the government, while the Socialist Party (PSI) enjoyed power only within centre-left coalitions.

The political configurations of Italian politics following the Second World War explain much about the long absence of discussions on family matters, and why social and political actors did not demand family policies. In Italy, family matters had been a source of tension and division between the main political parties ever since the drafting of the Constitution. After the approval of the Constitution, political parties avoided family issues because they highlighted the deep ideological divisions between Catholics and the Communist Party. Neither the Christian Democrat Party nor the Communist Party were interested in openly facing these issues. The DC and the Church avoided the issue because their vision of family relations adhered to fascist legal norms. In addition, the DC perceived social policy expansion as conflicting with the Church's interests in upholding existing state–family relations and preserving their role as social service providers, especially in the field of child care. The PCI, which approached these concerns as issues of women's rights separate from family arrangements, did not want to challenge Italian Catholic traditions and alienate its patriarchal working-class constituency.

In Spain, political allegiances on these issues were divided between centre right-wing parties supporting 'family unity' and left-wing parties supporting individual rights and 'women's policy'. The centre-right political parties in Spain were the only actors to advance demands for family policies in their political platforms and in parliamentary debates. However, the UCD in Spain (during the transition period from 1977 to 1982) did not advance proposals or push for approving a family policy, despite the intentions it stated in its political platforms. In this country the women's movement also added an important element of resistance to policies which seemed pro-natalist. Feminists in Spain, more than in other countries, identified social policies for family responsibilities (i.e. family allowances, tax deductions) as conservative and anti-feminist policies intended to uphold the traditional family model. They effectively conceptualized 'family' policy as inherently regressive while proposing their own alternative 'women's policy'.

Despite different political configurations, both countries exhibit similarities in the ideological representation of the family and of women's and men's roles within the family. In both countries the political parties' demands for social policies for families with children have been constrained by the ideological division over the representation of which family model should be supported by the state. In these two countries the left-oriented parties have avoided putting family policy on the political agenda and have preferred to frame requests in terms of social rights for women and individual rights.

The cultural and political hegemony exercised by the Catholic Church is fundamental for understanding Spain's and Italy's similar cultures of the family and the absence of demands for social policies for families. In addition, the cultural traits of the Italian and Spanish family systems, in particular the strength of intergenerational ties and kin solidarity, have been grounded on the enduring norms of rural society and patriarchal relations. These norms were reinforced during the authoritarian period. Family cultures stemmed from a cultural and social context largely oriented to Catholic values.

The strong cultural hegemony exercised by the Catholic Church on family definitions and on family issues hindered the suggestion of alternative definitions, delayed the reform of traditional family law (in Italy), and influenced divorce law in both countries. In addition, in Italy the Catholic Church also strongly influenced political life through more than 40 uninterrupted years of DC-led governments. Like the Italian DC, Church-connected associations and organizations in Spain have also avoided family policies up until very recently.

The Catholic vision of family relations remained the predominant legal family model up to the mid-1970s in both Italy and Spain. In addition, in both countries the Church exercised a negative influence on the development of public child care services, kindergartens and services for the elderly. At the same time, the primacy of the family (women) in caring work before other agencies was consistent with the principle of subsidiarity. This has also shaped the way in which these two states have developed poor relief in the 1990s – the system of *prestaciones no contributiva* in Spain and the *assegno al nucleo familiare* in Italy._

Despite these similarities, there are also important differences between the two national cases. In particular, the different role in demanding social policies for families and the different strength of social movements in the two countries must be underlined. Italy experienced important legal changes and welfare state expansion related to family relations and caring work during the 1970s. In the Italian social and political context at the end of the 1960s, demands for changing the institutional definition of the family model were framed mainly in terms of individual rights (women's rights) and/or child welfare, rather than in terms of family issues. These changes were initiated and strongly supported by social movements (feminist movements and trade unions). In Italy, as in other European countries, the feminist movement and women within the trade unions were especially active in the hard struggle for policies removing gender inequality and providing public child care services.

Soon after the end of the dictatorship, Spain experienced profound legal changes in family relations. Feminist movements and the Socialist Party (PSOE) framed their demands in terms of individual rights (and women's rights) or in terms of reducing social inequalities. But among the Spanish feminists, demands were framed in terms of 'women's policy'. They opted for a model of women's emancipation which implied a revolution in the domestic sphere (i.e. an equal division of family work between men and women). It did not require the support of state policies, such as public child care. This explains why feminists in Spain have hardly advanced any demands for public support for the care of children up until recently. Spanish trade unions were more concerned with pension reform and have remained practically silent on this issue.

The Weakness of Social and Political Actors

Through the four policy studies it has been possible to understand why social and political demands for social policies aimed at supporting the cost of children have been weak. Political and public discourse on the cost of children, when it existed, was obscured within debates on policies having different purposes. In particular, the analysis displays the incapacity and/or unwillingness of political actors to reckon with a public discussion about the model of social policy, about family models, and about the division of tasks between the state and the family.

The cases of family allowances in Italy and Spain and tax deductions in Spain show that in both countries, despite the different cultural and socio-economic contexts, the main social and political actors could hardly get rid of the unsolved 'family matter'. They were unable to overcome the ideological divisions about what family model they wanted. The latest reforms show how in these countries the issue of the cost of children has been quickly sacrificed to demands to reduce social expenditure and rationalize the social security system.

Social and political actors have little incentive to formulate strong demands for economic support to families with children. The Church has not been interested in having social policies for families in Italy or Spain. Trade unions played an ambiguous role in Italy. On the one hand, they have contributed to improving the amount of family allowances for quite some time; on the other hand, they have contributed to the maintenance of a fragmented and non-universal system of family allowances. Finally, trade unions in Italy have concentrated more resources and attention on issues which consolidate internal consensus, such as pension reform. In Spain the study of family allowance policy shows that the deep ideological divisions between the feminist movement and the PSOE on one hand, and the centre-right parties on the other, explain why family policies did not reach the Spanish political agenda for 15 years. The former regarded all family support measures as disguised attempts to impose a patriarchal model of the family, while the centre-right parties in general strove to support and defend the unity of just such a traditional family model. In addition to these deep ideological divisions, family reform was also delayed because of the need to form democratic consensus in a period of economic crisis.

The study of the Italian *nidi* policy shows that in the 1970s the presence of strong women and of organized working women was an important driving force in achieving the law. This demonstrates the importance of social pressure exercised from outside political parties for realizing certain social policies. However, *consociativismo,* or Italy's process of law-making through consensus, produced a law which in principle deviated from the original demands to support working women. Instead, it generated a law which reaffirmed the supremacy of the family and mother's care over public services. Therefore the *asilo nido* law did little to modify the cultural attitude that caring work should remain in the family (woman's) domain.

Opening the Black Box of the Family to Face Future Scenarios

Two main theoretical implications have emerged from this book. First, the analysis of the Mediterranean welfare state model should be revised to include not only the specific relationship between the family, the labour market and the state, but also the specific systems of family functioning and family relations. In fact, this work illustrates that the definition of kinship relations is an important dimension, together with gender and intergenerational relations, of the definition of the family model. These definitions are a consequence of the 'kin' notion of family relations included in the welfare state institutions, and are grounded on a given system of family functioning. Together, they illustrate one of the most unique mechanisms by which Mediterranean welfare states function.

The second theoretical implication is that the analysis of welfare states from a gender perspective should be revised in order to include not only gender relations, but also intergenerational and kin relations *within* the family. When these two cases are treated only from a gender perspective, they appear to match the family definitions found by other scholars for other countries. However, important historical and cross-national differences emerge with the introduction of intergenerational and kin dimensions. It becomes clear that family responsibility and the corresponding dependent family relationships have not been defined in the same ways across countries and over time. The boundary between state and family responsibility for reproduction and caring work has been redrawn over time and across countries. Knowledge about who within the family is responsible for caring work, and who is counted responsible for dependants, helps us to understand better both the functioning of the mixed economy of welfare states as well as the allocation of paid and unpaid work.

In Italy and Spain caring work is, like in other countries, largely performed by women. But what about the future? In Italy and Spain grandmothers or other unemployed or irregularly employed women have helped assure the continuous career patterns of other female relatives. Will this reproduction pattern remain the same in the future, given both the growing intergenerational imbalance and the increasing participation of women in the labour market?

To open the black box of the family and examine how family relations and power relations have been drawn and redrawn means facing future scenarios of private and public forms of welfare provision, and how they may or may not be mixed. In Italy and Spain the challenges to the functioning of the welfare state seem to be even greater than in other countries because of the sharp decline in birth rates, the growth of the aged population, the diminishing likelihood that working women will be able (and willing) to perform unpaid caring assistance for their children and their relatives, and the projected growth of the population in need of care and financial support given the reduction of the public social safety net. Indeed, all of these factors could undermine the main pillars of the Mediterranean model of the welfare state.

Yet, the recent rediscovery of the family in public and political debates in both countries stems neither from new demands for a more equal redistribution of paid and unpaid work, nor from a push to put legal norms and social policies in line with recent changes in the family. The re-emergence of the family in public debate comes instead from the need to find solutions to the welfare state crisis and the dramatic decline of fertility. Long-lasting ideological divisions and conflict over the family, as well as the reticence to open public debates on family changes and to legally and socially acknowledge new family patterns (cohabiting couples, single-person households, single parents' step-families, and so on), seem to remain unchanged. Any legal change altering the traditional definitions of the family, and any social policy rearranging the current balance of family responsibilities, continue to be perceived as attacks on 'the family'. In Spain and Italy it is the idea of the family, rather than the family/ies that each of us experiences and perceives in daily life, that remains the object of definition and the place that every social and political force wants to defend.

APPENDICES

Appendix 1
LEAVE SCHEMES FOR PARENTS IN EU COUNTRIES AT 31 DECEMBER 1993

	Maternity	**Paternity**	**Parental**
Belgium	15 weeks. General right for all female employees. Paid: earnings-related. Self-employed women entitled to 3 weeks postnatal leave, with a lump sum maternity grant.	3 days at 100% of wages.	Public sector workers can take a maximum of 3 months unpaid leave, to be taken immediately after maternity leave. Private sector employees (either parent) can receive a favourable level of unemployment benefits if they stop work or work part-time to care for a child. This counts as full-time work for social security if taken within 6 months to 3 years after the child's birth.
Denmark	18 weeks. General right for all employed and self-employed women. Paid: earnings-related.	10 days, paid as for maternity leave.	10 weeks per family paid as for maternity leave.
Germany	14 weeks. General right for all female employees. Paid: earnings-related.	None	None
Greece	14 weeks. All insured women. Paid: earnings-related.	None	Up to 3 months unpaid for each parent; 6 months for lone parents. Cannot be transferred between parents. Employers can refuse if more than 8% of workforce on leave.
Spain	16 weeks. For women with 6 months contributions in previous year. Paid: earnings-related (up to a ceiling).	2 days at 100% wages. Mother can choose to transfer up to 4 weeks maternity leave to father.	12 months unpaid for family. Further 2 years available, but employee loses right to automatic reinstatement. Parents with children under 6 or disabled can also reduce their working hours by $^1/_3$ or $^1/_2$ but with no compensation for lost pay.

Appendix 1 (Continued)

	Maternity	Paternity	Parental
France	16 weeks (26 for the third and subsequent child). Women insured for at least 10 months for 1,200 hours in previous year. Paid: earnings-related.	3 days, to be taken during 15 days before and after birth.	For the first 3 years of child's life. Can be taken by either parent, shared or alternated. Unpaid, but for second and subsequent child can receive *Allocation Parental d'Educacion.*
Ireland	14 weeks +optional 13 unpaid. 39 weeks of insured employment in previous year or tax year. Paid: earnings-related (with minimum and maximum).	None	None
Italy	22 weeks. General right for all employed and self-employed women. Paid: earnings-related. During first year after birth mothers in full-time work can take 1 hour rest period per day (or shorten day).	None	6 months after maternity leave. Must be taken before child's 1st birthday. Mainly for mother but can be partly transferred to father. Paid at 30% of wages. Leave can be extended to first 3 years of life, paid at 30% of earnings, or can be taken as 2 hours per day paid leave when child is seriously disabled.
Luxembourg	16 weeks. 6 months of contributions. Paid: earnings-related.	None	None
Netherlands	16 weeks. General rights for all employed women. Paid: earnings-related.	None	6-month period for each parent when they can work reduced hours (to a minimum of 20 hours per week). Can be taken at any time until a child is 4 years old. Entitlement is not transferable between parents and there is no compensation for lost wages except for lone parents whose pay falls below social assistance level.

Appendix 1 (Continued)

	Maternity	**Paternity**	**Parental**
Portugal	90 days. All insured women. Paid: earnings-related.	None	6–24 months per family, to be taken after maternity leave. Unpaid. Where a child is under 12 or disabled a (non-managerial) parent can work half their normal hours, but without compensation for lost pay.
United Kingdom	40 weeks: 6 weeks paid: earnings-related; 12 weeks flat rate and remainder unpaid. Does not carry continuous pension entitlement. 2 years full-time or 5 years part-time insured employment with same employer.	None	None

Source: Ditch and Barnes (1996).

Appendix 2

SURVIVORS' PENSION SCHEMES IN BELGIUM, FRANCE, GERMANY, ITALY AND SPAIN, 1995

	Widow/widower	Widower	Children	Other beneficiaries
Belgium	Surviving spouse: spouse married to the deceased for at least 1 year. Aged at least 45 or bringing up a child or being an invalid. Economically inactive. Benefits: 80% of the actual or hypothetical retirement pension, of the insured person calculated at the rate for a married couple where the spouse is dependent. Minimum ECU 8,316 per year.	Since 1983	No orphan's pension, but special family allowance scheme.	None
France	Surviving spouse: for revision pension: widow or widower (with insufficient means of existence), aged 55, of a beneficiary of old-age pension. For invalid widow's pension and and widow's (widower's) old-age pension: Widow or widower aged 55 or above and disabled. Benefits: 54% of real or hypothetical old-age pension of the deceased person. Minimum ECU 2,549 per year.	Since 1978	Survivors' pension when surviving spouse has at least the charge of one child under 16. Increase ECU 73 per month and per child.	None
Germany	Surviving spouse: married to the deceased at the time of his/her death, or divorced before 1 July 1977 and financially dependent upon the deceased. The surviving partner must not have married again. Benefits: 60% of the deceased spouse's pension rights if the widow/widower is above age of 45, caring for children or unfit to work. Otherwise 25%. Own income above a certain threshold reduces the pension by 40% of the exceeded amount.	Since 1985	1/10 of insured person's pension plus children's supplement. No restriction on combination with family allowance. Age limit: 18 years (27 for study or occupational training).	None

Appendix 2 (Continued)

	Widow/widower	Widower	Children	Other beneficiaries
Italy	Surviving spouse: widow or widower. In case of divorce, a widow/widower receiving maintenance can obtain the survivor's pension at the discretion of a judge. Benefits: 60% of the insured person's invalidity or old-age pension. Since 1995 a degree of income testing has been introduced. Thus, survivor's pension can be reduced from 50% to 25% if the surviving spouse has a pension at least three times higher than the minimum pension level.	Since 1977	Under 18 or not able to work. Under 21 or 26 for study's reasons. From 20 to 40% each child.	For parents, brothers or sisters (15% of the insured person's pension) provided there are no other survivors and that they were dependants of the deceased when death occurred.
Spain	Surviving spouse: widow and widower: must have lived with the deceased insured on a regular basis. In case of separation or divorce pension is shared between beneficiaries in proportion to the length of period of cohabitation. Benefits: 45% of reference wage for deceased person. Minimum pension over 65 ECU 314 per month. Between 60 and 65 ECU 274 per month. Under 60, 209 ECU per month.	Since 1983	Under 18	Pension (under certain conditions) 20% of the reference figure for grand-children, siblings, mother and father, grandmothers and grandfathers.

Source: *MISSOC (1996).*

Appendix 3
FAMILY AND TAXATION IN SOME SELECTED COUNTRIES, 1990

	Unit of assessment	**Dependent spouse**	**Dependent children**	**Other family dependent members**
Italy	Individual since 1977	Tax credits	Tax credits	Tax credits
Germany	Individual/family (optional)	Income-splitting	Tax allowances	No
France	Family	*Quotient familial*	*Quotient familial*	No
Spain	Individual/family (optional) since 1988	Tax credits	Tax credits	Tax credits
Belgium	Individual since 1989	Tax credits	Both spouses receive married person's allowance	No

Source: OECD (1991).

Appendix 4
SOCIAL EXPENDITURES ON THE FAMILY, AS PERCENTAGE OF TOTAL SOCIAL EXPENDITURE IN THE EU COUNTRIES, 1980 AND 1991

Countries	**1980**	**1991**
Belgium	10.7	8.0
Denmark	9.9	10.3
France	10.7	8.2
Germany	9.0	6.0
Greece	3.6	1.4
Ireland	8.0	10.6
Italy	6.9	3.6
Luxembourg	8.8	9.5
Netherlands	8.9	5.5
Portugal	6.8	5.4
Spain	2.7	0.6
UK	11.4	9.6
EC 12	9.2	6.4

Source: Commission of European Communities (1993).

Appendix 5

THE TAX AND SOCIAL SECURITY TREATMENT OF CHILDREN, 1991[a]

Countries	Amount of relief/credit		Family allowances	
	2 children	3 children	2 children	3 children
Belgium	12.4	28.1	13.7	25.7
Germany	13.4	20.0	3.0	5.8
Spain	0.3	0.4	0.4	0.6
Italy	1.0*	1.5*	7.1	10.6
France**	7.1	18.2		

Source: Sterdyniak (1994), table 1, p. 370.

Notes:

a The values are expressed as percentage of the average blue-collar worker's wage.

* For Italy the figure is divided by two if the wife works.

** France applies the *quotient familial.*

Appendix 6

NET INCOME EARNED BY TWO-CHILD WORKING FAMILIES, AFTER TAXES AND SOCIAL SECURITY CONTRIBUTIONS, IN COMPARISON WITH SINGLE PEOPLE*

Year	1981	1984	1987	1988	1989	1990
Belgium	18.2	19.4	19.9	21.5	25.6	24.5
France	13.4	15.2	14.2	14.2	14.0	14.4
Germany	12.8	11.8	14.5	13.7	13.5	13.7
Italy	11.1	13.1	7.0	11.5	10.9	10.7
Spain	5.9	5.6	4.9	5.7	5.6	5.5

Source: OECD, Tax/Benefit Position of Production Workers, Paris, 1986, 1991, table 4.

Note: * Net income as percentage of gross earnings.

Appendix 7
CHILD BENEFIT PACKAGE: COUPLES WITH CHILDREN, BEFORE AND AFTER HOUSING, 1994

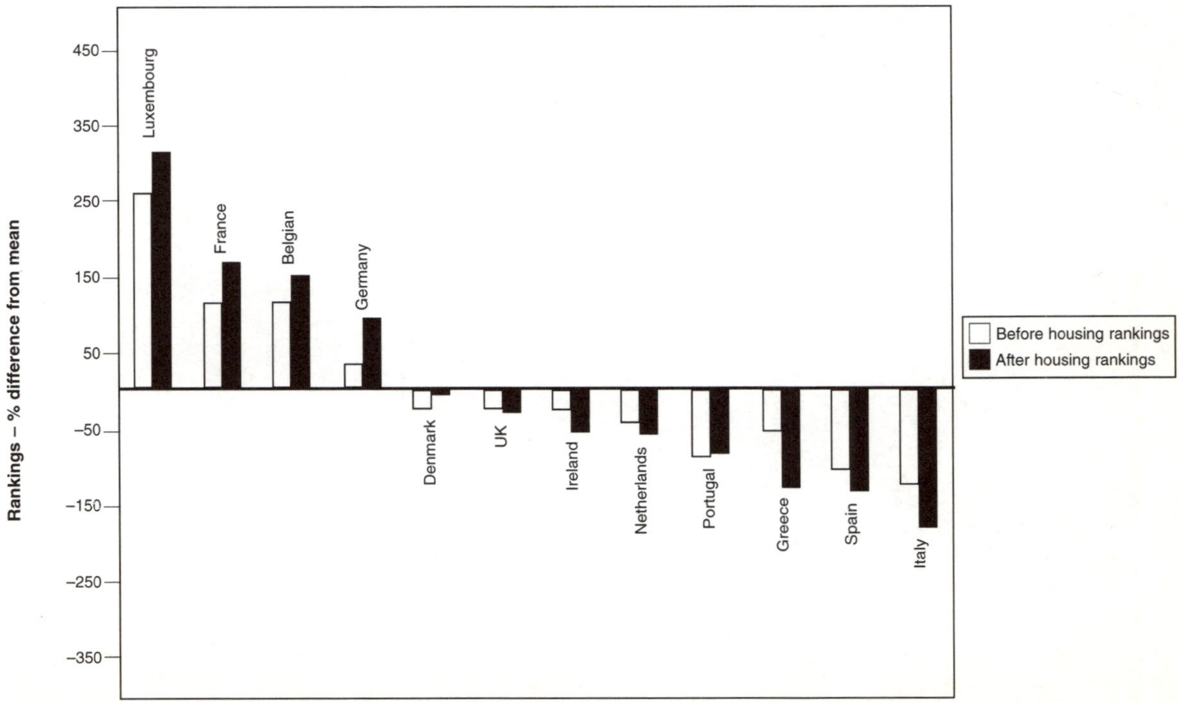

Source: Ditch and Barnes *et al.* (1996), table 5.5a/b, p. 84.

Appendix 8
PUBLIC PROVISION OF CHILD CARE, 1988*

Country	Year	Children under 3	Age 3 to school age
Belgium	1988	20	95+
Denmark	1989	48	85
France	1988	20	95+
Germany	1987	3	65–70
Greece	1988	4	65–70
Ireland	1988	2	55
Italy	1986	5	85+
Luxembourg	1989	2	55–60
Netherlands	1989	2	50–55
Portugal	1988	6	35
Spain	1988	na	65–70
United Kingdom	1988	2	35–40

Source: Gauthier (1996), p. 181. na = not available

Note: *All figures in percentages.

REFERENCES

Alber, J. (1995) 'A Framework for the Comparative Study of Social Services', *Journal of European Social Policy*, 5.2, pp. 131–49.

Alberdi, I. (1992) 'Cambios en el derecho de la familia y su repercusiones sociales', *Infancia y Sociedad,* 16, pp. 35–48.

Alberdi, I. (1994) 'Organización formal, marco legal de la familia', in C. Alberdi, *et al.* (eds), *Informe sobre la situación de la familia en España* (Madrid: Ministerio de Asuntos Sociales).

Alberdi, I. (1997) 'La familia. Convergencia y divergencia de los modelos familiare Españoles en el entorno Europeo', unpublished paper (Madrid: Universidad Complutense).

Alberdi, I., Flaquer, L. and de Ussel, J. (1994) *Parejas y matrimonios: actitudes, comportamientos y experiencias* (Madrid: Ministerio de Asuntos Sociales).

Alcaide Inchausti, J. (1988) 'El gasto publico en la democracia Española. Los hechos', *Papeles de economía española*, 37.

Almeda, E. and Sarasa S. (1996) 'Spain: Growth to Diversity', in V. George and P. Taylor-Gooby (eds), *European Welfare Policy. Squaring the Welfare Circle* (London: Macmillan), pp. 155–76.

Alonso Olea, A. M. (1959) *Instituciones de seguridad social* (Madrid: Instituto de Estudios Políticos).

Alonso Olea, A. M. (1994) *Las Prestaciones Sanitarias de la Seguridad Social* (Madrid: Editorial Civitas, S.A.).

Ambrosini, M. and Rossi, G. (1994) 'Lavoro extradomestico delle donne, tendenze demografiche e politiche familiari', in G. Blangiardo, A. Golini *et al.* (eds), *Politiche per la popolazione in Italia* (Torino: Edizione della Fondazione Giovanni Agnelli).

Anttonen, A. and Sipilä, J. (1996) 'European Social Care Services: Is It Possible to Identify Models?', *Journal of European Social Policy*, 5.2, pp. 87–100.

AP (Alianza Popular) (1982) *Soluciones para España* (party programme).

AP (Alianza Popular) (1986) *Esto tiene arreglo* (party programme).

Ardigò, A. (1974) 'Funzioni e limiti del welfare state', *Ricerca Sociale*, 5, pp. 5–30.

Ascoli, U. (1984) *Welfare state all'Italiana* (Bari: Laterza).

Auletta, T. A. (1984) *Alimenti e solidarietà familiare* (Milan: Giuffrè Editore).

Balbo, L. (1976) *Stato di famiglia. Bisogni, privato, collettivo* (Milan: Etas Kompass).

Balbo, L. (1978) 'La doppia presenza', *Inchiesta*, 32, pp. 3–6.

Balbo, L. (1984) 'La complessa società dei servizi', *Inchiesta*, 66 (Oct.–Dec.), pp. 1–7.

Balbo, L. (1987) 'Family, Women, and the State: Notes Toward a Typology of Family Roles and Public Intervention', in C. S. Maier (ed.), *Changing Boundaries of the Political* (Cambridge: Cambridge University Press).

Baldwin, P. (1990) *The Politics of Social Solidarity* (Cambridge: Cambridge University Press).

Ballestrero, M. V. (1979) *Dalla tutela alla parità* (Bologna: Il Mulino).

Ballestrero, M. V. (1989) *Oltre la parità. Parità, pari opportunità, azioni positive* (Rome: Ediesse).

Ballestrero, M. V. (1996) 'La protezione concessa e l'uguaglianza negata: il lavoro femminile nella legislazione italiana', in A. Groppi (ed.), *Il lavoro delle donne* (Bari: Editori Laterza).

Bar, A. (1984) 'The Emerging Spanish Party System: Is There a Model?', in S. Bartolini and P. Mair (eds), *Party Politics in Contemporary Western Europe* (London: Frank Cass).

Barbagli, M. (1988) *Sotto lo stesso tetto* (Bologna: Il Mulino).

Barbagli, M. (1990) *Provando e riprovando. Matrimonio e divorzio in Italia e negli altri paesi occidentali* (Bologna: Il Mulino).

Barbagli, M. (1996) 'Asymmetry in Intergenerational Family Relationships in Italy', in T. Hareven (ed.), *Ageing and Generational Relations Over the Life Course* (Berlin: Walter de Gruyter).

Barbagli, M. and Kertzer, D. I. (1992) *Storia della famiglia italiana* (Bologna: Il Mulino).

Barbagli, M. and Saraceno, C. (eds) (1997) *Lo stato delle famiglie in Italia* (Bologna: Il Mulino).

Barea, J. and Gonzales-Pàramo, J. M. (1996) *Pensiones y prestaciones por desempleo* (Madrid: Fundacion BBV).

Bartolini, S. and Mair, P. (1990) *Identity, Competition, and Electoral Availability* (Cambridge: Cambridge University Press).

Bartolini, S. (1993) 'On Time and Comparative Research', *Journal of Theoretical Politics*, 5.2, pp. 131–67.

Bautista Parejo, E. (1996) 'Mujer y democracia en España. Evolución jurídica y realidad social', *Revista de Estudios Sociales y Sociología Aplicada*, 105 (Oct.–Dec.), pp. 49–71.

Bazo, M. T. (1990) *La sociedad anciana* (Madrid: CIS).

Berea, R. (ed.) (1992) *Analisi de la demanda de servicios para la primera infancia* (Madrid: Ministerio de Asuntos Sociales).

Bertelsmann Foundation (ed.) (1999) *International Reform Monitor*, issue 1 (Gütersloh: Bertelsmann Foundation). http://www.reformmonitor.org)

Bertelsmann Foundation (ed.) (2000a) *International Reform Monitor*, issue 2 (Gütersloh: Bertelsmann Foundation). http://www.reformmonitor.org)

Bertelsmann Foundation (ed.) (2000b) *International Reform Monitor*, issue 3 (Gütersloh: Bertelsmann Foundation). http://www.reformmonitor.org)

Bettio, F. (1988) *The Sexual Division of Labour* (Oxford: Oxford University Press).

Bettio, F. and Villa, P. (2000) 'To What Extent Does it Pay to be Better Educated? Education and Market Work for Women in Italy', in M. J. Gonzalez, T. Jurado and M. Naldini (eds), *Gender Inequalities in Southern Europe: Women, Work and Welfare in the 1990s* (London: Frank Cass), pp.150–70.

Bianchi, M. (1981) *I servizi sociali* (Milan: Angeli).

Bimbi, F. (1978) 'L'ideologia cattolica tra crisi e ristratturazione della famiglia', in G. Guizzardi (ed.), *La politica dell'ideologia* (Padova: Cleup).

Bimbi, F. (1989) 'The Double Presence. A Complex Model of Italian Women Labour', *Marriage and Family Review*, 1.2, pp. 81–105.

Bimbi, F. (1992) 'La cittadinanza delle donne', *Inchiesta* (July–Dec.).

Bimbi, F. (1993) 'Gender, Gift Relationship and the Welfare State Culture in Italy', in J. Lewis (ed.), *Women and Social Policies in Europe. Work, Family and the State* (London: Edward Elgar), pp.138–69.

Bimbi, F. (2000) 'The Family Paradigm in the Italian Welfare State (1947–1996)', in M. J. Gonzalez, T. Jurado and M. Naldini (eds), *Gender Inequalities in Southern Europe: Women, Work and Welfare in the 1990s* (London: Frank Cass), pp. 72–88.

Bimbi, F. and Della Sala, V. (1998) 'L'Italie. Concertation sans représentation', in J. Jenson and M. Sineau (eds), *Qui doit garder le jeune enfant*? (Paris: Droit et Société LGDJ).

Bison, I., Pisati, M. and Schizzerotto, A. (1996) 'Disuguaglianze di genere e storie lavorative', in S. Stella and C. Saraceno (eds), *Genere. La costruzione sociale del femminile e del maschile* (Bologna: Il Mulino), pp. 253–79.

Bock, G. and Thane, P. (1991) *Maternity and Gender Policies* (London: Routledge).

Bonini, F. (1997) 'La famiglia alla costitutente. Strategie e modelli istituzionali', in F. Mazzonis (ed.), *Percorsi e modelli familiari in Italia tra '700 e '900* (Rome: Bulzoni), pp. 207–54.

Borrajo Iniesta, S. (1987) 'Estudio sociológico sobre la ruptura matrimonial en Madrid capital (1981–84)', *Revista Española de Investigaciones Sociológicas*, 37, pp. 113–37.

Borrajo Iniesta, S. (1990) *La ruptura matrimonial en España (*Madrid: Eudema).

Bradshaw, J. and Ditch, J. (1993) 'Support for Children. A Comparison of Arrangements in Fifteen Countries', *Research Report*, 21 (London: HMSO).

Bruckner, E., Knaup, K. and Müller, W. (1993) 'Soziale Beziehungen und Hilfeleistungen in Modernen Gesellschaften', working paper, Mannheim Centre for European Social Research.

Bussemaker, J. and Van Kersbergen, K. (1994) 'Gender and Welfare States: Some Theoretical Reflections', in D. Sainsbury (ed.), *Gendering Welfare States* (London: Sage), pp. 8–26.

Buttafuoco, A. (1991) 'Motherhood as a Political Strategy: The Role of the Italian Women's Movement in the Creation of the *Cassa Nazionale di Maternita*', in G. Bock and P. Thane (eds), *Maternity and Gender Policies* (London: Routledge).

Cafagna, L. (1961) 'L'industrializzazione Italiana. La formazione di una base industriale fra il 1896 e il 1914', *Studi Storici*, 7, pp. 652–96.

Caldwell, L. (1991) *Italian Family Matters: Women, Politics and Legal Reform* (London: Macmillan).

Carr, R. and Fusi, J. P. (1987) *Spain: Dictatorship to Democracy* (London: Allen and Unwin).

Carreras, A. (1989) *Estadisticas Historicas de Espana. Siglo XIX–XX* (Madrid: Fundación Banco Exterior).

Castles, F. (1978) *The Social Democratic Image of Society* (London: Routledge and Kegan Paul).

Castles, F. (1994) 'Social Security in Southern Europe: A Comparative Overview', paper presented at the conference organized by the subcommittee on *Southern Europe of the Social Science Council*, Istanbul, 1–2 July.

Cavalli, A., *et al.* (1993) *Terzo rapporto IARD sulla condizione giovanile in Italia* (Bologna: Il Mulino).

Cazzola, F., *et al.* (1975) *Il decreto legge fra Governo e Parlamento* (Milan: Giuffrè Editore).

CD (Coalición Democratica) (1979) *Porque las cosas no están centradas. ¡Ahora! Vota Coalición Democrática* (party programme).

Censis/Cnel (1980–97) *XIV–XXII Rapporto sulla situazione sociale del Paese* (Rome: Franco Angeli).

CGIL (Camera Generale Italiana Lavoro) (1981) 'Mozione sulla politica fiscale', *Rassegna Sindacale,* 43, pp. 160–3.

Cherubini, A. (1977) *Storia della previdenza sociale in Italia* (Rome: Editori Riuniti).

CIPE (Commissione di indagine sulla povertà e l'emarginazione) (1995) *Verso una politica di lotta alla povertà. L'assegno per i figli e il minimo vitale* (Rome: Presidenza del Consiglio dei Ministri).

Cires (1993) *La realidad social en España 1991–1992* (Barcelona: Ediciones B.).

CIS (Centro de Investigaciones Sociológicas) (1983) 'Educación Preescolar: Madres', *estudio* n. 1373 (Madrid: CIS).

CNAF (Caisse Nationale des Allocations Familiales) (1995) 'Concept et contextes dans les comparaisons internationale une observation sur les politiques familiales en Europe 1993–95', *Rapport final à la CNAF*, Comparisons des politiques familiales en Europe co-financé par ESRC et la Commision Européenne (November).

Coll, P. and Martín, R. (1989) 'La protección de la maternidad, de la familia y de la vivienda', *Revista de Economía y Sociología del Trabajo*, 3, pp. 67–83.

Colozzi, I., and Matteini, M. (1991) 'La promozione sociale della famiglia: proposte legislative ed esperienze di servizi innovativi', in P. Donati (ed.), *Secondo rapporto sulla famiglia in Italia* (Milan: Edizioni San Paolo), pp. 446–514.

Comitato per l'anno internazionale della famiglia (1994) *Per una politica familiare in Italia* (Rome: Presidenza del Consiglio).

Commission of the European Community (1988) *Young Europeans in 1987* (Luxembourg: CEC).

Commission of the European Community (1992) *Die Lage der Frauen auf dem Arbeitsmarkt. Entwicklungstendenzen in den zwölf Ländern der Europäischen Gemeinschaft 1983–1990,* Die Sonderhefte von Frauen Europas, 36.

Commission of the European Community (1993) *Social Expenditure and Receipts. 1980–1991* (Luxembourg: CEC).

Cotarelo, R. (1992) 'Los partidos politicos', in R. Cotarelo (ed.), *Transicion politica y consolidacion democratica en España (1975–1986)* (Madrid: CIS), pp. 299–326.

Council of Europe (1990) 'Household Structures in Europe', *Population Studies*, 22.

Cousins, C. (1995) 'Women and Social Policy in Spain: The Development of a Gendered Welfare Regime', *Journal of European Social Policy*, 5.3, pp. 175–97.

Cruz Roche, I. (1984) *Análisis económico de la seguridad social Española: 1972–1982* (Madrid: Ministerio de Trabajo y Seguridad Social).

Cuesta Bustillo, J. (1988) *Los seguros sociales en la España del siglo XX. Hacia los seguros sociales obligatorios. La crisis de la Restauración* (Madrid: Ministerio de Trabajo y Seguridad Social).

Cuesta Bustillo, J. (1994) 'Hacia el seguro de maternidad: la situación de la mujer obrera en los años veinte', in Instituto Universitario de Estudios de la Mujer (ed.), *Ordinamento Juridico y realidad social de la mujeres, siglos XVI a XX,* Actas de las cuartas jornadas de investigación interdisciplinaria (Madrid: University of Madrid), pp. 321–55.

Daly, M. (1994) 'Comparing Welfare States: Towards a Gender Friendly Approach', in D. Sainsbury (ed.), *Gendering Welfare States* (London: Sage).

Daly, M. (2000) *The Gender Division of Welfare* (Cambridge: Cambridge University Press).

David, P. (1984) 'Il sistema assistenziale in Italia', in U. Ascoli (ed.), *Welfare state all'italiana* (Bari: Laterza), pp. 185–205.

DC (Democrazia Cristiana) (1968) 'Il programma della DC al servizio del Paese', *Il Popolo*, 7, p. 4.

DC (Democrazia Cristiana) (1987) *Un programma per l'Italia. Elezioni politiche 14–15 giugno 1987* (party programme).

DC (Democrazia Cristiana) (1992) *Un programma per l'Italia verso l'Europa. Prima l'Italia. Libertas, fai vincere il tuo futuro* (party programme).

De Miguel, A. (1992) *La sociedad Española, 1992–1993* (Madrid: Alianza Editorial).

De Rosa, L. (1980) *La rivoluzione industriale in Italia* (Rome: Laterza).

Del Re, A. (1995) 'Social and Family Policies in Italy: Comparisons with France', *Cross-National Research Papers*, 18, pp. 68–79.

Delgado Pérez, M. and Livi-Bacci, M. (1992) 'Fertility in Italy and Spain: The Lowest in the World', *Family Planning Perspectives*, 24.4, pp. 162–71.

Delgado, M. (1993) 'Cambios recientes en el proceso de formación de la familia', *Revista Española de Investigaciones Sociales*, 64, pp. 123–53.

Di Febo, G. and Saba, M. (1994) *Ordinamento jurídico y realidad social de las mujeres siglo XVI a XX*, Actas de las cuartas jornadas de investigaciòn interdisciplinaria (Madrid: University of Madrid), pp. 439–52.

Diez-Pícazo, L. and Gullón, A. (1990) *Sistema de derecho civil*, vol. 4 (Madrid: Tecnos).

Ditch, J., Barnes, H. *et al.* (eds) (1996) *A Synthesis of National Family Policies 1994*, European Observatory on National Family Policy (Brussels: Commission of the European Community).

Donati, P. (1981) *Famiglia e politiche sociali* (Milan: Angeli).

Donati, P. (1984) *Risposte alla crisi dello stato sociale* (Milan: Angeli).

Donati, P. (1989) *Primo rapporto sulla famiglia in Italia* (Milan: Angeli).

Donzelot, J. (1977) *Police de la famille* (Paris: Minuit).

Dopffel, P. (1988) 'Child Support in Europe: A Comparative Overview', in A. Kahn and S. Kamerman (eds), *Child Support From Debt Collection to Social Policy* (London: Sage), pp. 176–223.

Eisenstein, Z. (1979) 'Developing a Theory of Capitalist Patriarchy and Socialist Feminist', in Z.

Eisenstein (ed.), *Capitalist Patriarchy and the Case for Socialist Feminism* (New York: Monthly Review).
Escario, P., Alberdi, I.. *et al.* (eds) (1996) *Lo personal es político. El movimiento feminista en transición* (Madrid: Ministerio de Asuntos Sociales/Instituto de la Mujer).
Escobedo, A. (1999) 'Work–Family Arrangements in Spain: Family Adjustments to Labour Market Imbalances', in L. Den Dulk, A. Van Doorne-Huiskes and J. Schippers (eds), *Work–Family Arrangements in Europe* (Amsterdam: Thesis Publishers), pp. 103–29.
Espina, A. (1990) *Empleo, democracia y relaciones industriales en España. De la industrialización al mercado único* (Madrid: Ministerio de Trabajo y Seguridad Social).
Esping-Andersen, G. (1990) *The Three Worlds of Welfare Capitalism* (New York: Polity Press).
Esping-Andersen, G. (1996) 'Igualidad o Empleo? La interacción de salarios, estado de bienestar y cambio familiar', in *Dilemas del Estado de Bienstar* (Madrid: Fundación Argentaria), pp. 9–27.
Esping-Andersen, G. (1999) *Social Foundations of Postindustrial Economies* (Oxford: Oxford University Press).
Eurobarometer (1990) *European Public Opinion on the Family and the Desire for Children*, 32 (Brussels, Luxembourg).
Eurobarometer (1991a) *Employment and Family Within the Twelve*, 34 (Brussels, Luxembourg).
Eurobarometer (1991b) *Young Europeans*, 34.2 (Brussels, Luxembourg).
Eurobarometer (1992) data file obtained by *Zentrum für Europäische Umfrageanalysen und Studien*, 37.2, Mannheim.
Eurobarometer (1993) *Europeans and the Family*, 39.0 (Brussels, Luxembourg).
Eurostat (1984) *Employment and Unemployment* (Brussels, Luxembourg).
Eurostat (1991) *Labour Force Survey. Results 1989* (Brussels, Luxembourg).
Eurostat (1992a) *Employment and Unemployment* (Brussels, Luxembourg).
Eurostat (1992b) *Demographic Statistics 1992* (Brussels, Luxembourg).
Eurostat (1993) *Labour Force Survey. Results 1991* (Brussels, Luxembourg).
Eurostat (1995) *Statistiques Démographiques* (Brussels, Luxembourg).
Eurostat (1997) *Les Jeunes de l'Union Européenne* (Brussels, Luxembourg).
Eurydice (Information Network in Europe) (1996) *Key Data on Education in the European Union* (ECSC–EC–EAEC, Brussels, Luxembourg).
Fabiani, G. (1997) *L'agricoltura in Italia tra sviluppo e crisi 1945–1977* (Bologna: Il Mulino).
Facchini, C. (1997) 'Gli anziani e la solidarietà tra le generazioni', in M. Barbagli and C. Saraceno (eds), *Lo stato delle famiglie in Italia* (Bologna: Il Mulino) pp. 281–9.
Falcucci, F. (1967) 'Relazione al XI Convegno Nazionale del Movimento Femminile della Democrazia Cristiana', Bologna 23–25 September 1966 (Rome: Arti Grafiche Italiane).
Fargion, V. (1983) 'L'assistenza pubblica in Italia dall'unità al fascismo: Primi elementi per un'analisi strutturale', *Rivista Trimestrale di Scienza dell'Amministrazione*, 2, pp. 476–85.
Fargion, V. (1995) 'The Territorial Cleavage in Italy and the Welfare Mix in Social Services. Regional Policies in the 1980s', paper presented at the Comparative Research on Welfare State Reforms conference, Pavia, 14–17 September.
Fargion, V. (1998) *Geografia della cittadinanza sociale in Italia* (Bologna: Il Mulino).
Farneti, P. (1985) *The Italian Party System (1945–80)* (London: Pinter).
Federici, N. (1984) *Procreazione, famiglia, lavoro della donna* (Torino: Loescher Editore).
Ferrera, M. (1984) *Welfare state in Italia. Sviluppo e crisi in prospettiva comparata* (Bologna: Il Mulino).
Ferrera, M. (1989) 'The Welfare System Environment', in J. Dixon and R. P. Scheurell (eds), *Social Welfare in Developed Market Countries* (London: Routledge).
Ferrera, M. (1993) *Modelli di solidarietà. Politica e riforme sociali nelle democrazie* (Bologna: Il Mulino).
Ferrera, M. (1996) 'The Southern Model of Welfare in Social Europe', *Journal of European Social Policy*, 6.1, pp. 17–37.

Ferrera, M. (1996) 'Il modello di welfare sud Europeo', *Quaderni di ricerca*, 5 (Milan: Università Bocconi, Poleis).

Ferrucci, F. (1998) 'Gli orientamenti familiari in Italia alla fine degli anni 90', in P. Di Nicola (ed.) *Sociologia e politiche sociali*, 1.3, pp. 47–77.

Finch, J. (1989) *Family Obligations and Social Change* (Cambridge: Polity Press).

Flaquer, L. (1990) 'La familia Española: cambio y perspectivas', in S. Giner (ed.), *España. Sociedad y política* (Madrid: Espasa-Calpe).

Flaquer, L. (1994) 'La nupcialidad', in M. Juárez (ed.), *V Informe sociológico sobre la situación social en España. Sociedad para todos en el año 2000* (Madrid: FOESSA).

Flaquer, L. (1998) 'The Distribution of Tasks Between the Welfare State and the Family in Spain: In Search of a Comprehensive Explanation', paper presented at the Family and Family Policies in Southern Europe workshop, University of Torino, 20–21 November.

Flaquer, L. and Soler, J. (1990) *Permanencia y cambio en la familia española* (Madrid: CIS).

Flora, P. (1985) 'On the History and Current Problems of the Welfare State', in S. N. Eisenstad and O. Ahimeir (eds), *The Welfare State and its Aftermath* (London: Croom Helm).

Flora, P. (1986) 'Introduction', in P. Flora (ed.), *Growth to Limits*, vol. 1 (Badia Fiesolana: EUI).

Flora, P. and Heidenheimer, A. J. (eds) (1981a) *The Development of the Welfare State in Europe and America* (New Brunswick: Transaction).

Flora, P. and Heidenheimer, A. J. (1981b) 'The Historical Core and Changing Boundaries of the Welfare State', in P. Flora and A. J. Heidenheimer (eds), *The Development of the Welfare State in Europe and America* (New Brunswick: Transaction), pp. 17–36.

Flora, P., Kraus, F. *et al.* (eds) (1987) *State, Economy, and Society in Western Europe*, 1915–1975, vol. 2, *The Growth of Industrial Societies and Capitalist Economies* (London: Macmillan; Frankfurt: Campus Verlag).

Fox Harding, L. (1996) *Family State and Social Policy* (London: Macmillan).

Franco, D. (1993) *L'espansione della spesa pubblica in Italia* (Bologna: Il Mulino).

Franco, D. and Sartor, N. (1990) *Stato e famiglia* (Milano: F. Angeli).

Freire, J. M. (1993) 'Cobertura sanitaria y equidad en España', in *I Simposio sobre igualdad y distribución de la renta y la riqueza* (Madrid: Fundación Foessa), pp. 113–38.

Friot, B. (1997) *Puissances du salariat. Emploi et protection sociale à la française* (Paris: La Dispute).

Fua, G. (1980) *Problemi dello sviluppo tardivo in Europa* (Bologna: Il Mulino).

Gambale, S. (1994) 'Il trattamento tributario della famiglia', in Committee of the International Year of the Family, *Per una politica familiare in Italia* (Rome: Presidenza del Consiglio dei Ministri), pp. 91–108.

Garrido Medina, L. (1993) 'La familia estatal: El control fiscal de la natalidad', in L. Garrido Medina and E. Gil Calvo (eds), *Estrategias familiares* (Madrid: Alianza Universitaria).

Gauthier, A. H. (1991) 'Family Policies in Comparative Perspective', discussion paper 5, Centre for European Studies (Oxford: Nuffield College).

Gauthier, A. H. (1996) *The State and the Family. A Comparative Analysis of Family Policies in Industrialized Countries* (Oxford and New York: Oxford University Press).

Gauthier, A. H. (1999) 'The Sources and Methods of Comparative Family Research', *Comparative Social Research*, 18, special issue, 'Family Change: Practices, Policies and Values', ed. A. Leira, pp. 31–56.

Generalitat de Catalunya (1993) *Pla integral de suport a les families* (Barcelona: Department de Benestar Social).

Giddens, A. (1979) *Central Problems in Social Theory* (London: Macmillan).

Giner, S. (1985) 'Political Economy, Legitimation and the State in Southern Europe', in R. Hudson and J. Lewis (eds), *Uneven Development in Southern Europe* (London and New York: Methuen).

Giner, S. (ed.) (1990) *Espana sociedad y politica* (Madrid: Espasa-Calpe).

Giner, S. and Pérez Yruela, M. (1979) *La sociedad corporativa* (Madrid: Centro de Investigaciones Sociológicas).

Giner, S. and Sarasa, S. (1992) 'Religion, politica y modernidad en España', *Revista International de Sociología*, Tercera Epoca, 1, pp. 9–60.

Ginsborg, P. (1989) *Storia d'Italia dal dopoguerra a oggi. Società e politica 1943–1988* (Torino: Einaudi).

Ginsburg, N. (1992) *Divisions of Welfare* (London: Sage).

Golini, A. (1988) 'Profilo demografico della famiglia italiana', in P. Melograni (ed.), *La famiglia italiana dall'ottocento ad oggi* (Bari: Laterza), pp. 327–81.

Golini, A. and Menniti, A. (1994) 'Popolazione e famiglie', in Committee of the International Year of the Family, *Per una Politica Familiare in Italia* (Rome: Presidenza del consiglio dei ministri), pp. 3–26.

González, M. J., Jurado, T. and Naldini, M. (2000) 'Interpreting the Transformation of Gender Inequalities in Southern Europe', in M. J. González, T. Jurado and M. Naldini (eds), *Gender Inequalities in Southern Europe. Women, Work and Welfare in the 1990s* (London: Frank Cass), pp. 4–34.

González-Sancho Lopez, E. (1982) 'La politica de proteccion a la familia en España', *Papeles De Economia Española*, 12/13, La Seguridad Social.

Goode, W. (1993) *World Changes in Divorce Patterns* (New Haven and London: Yale University Press).

Gorrieri, E. (1979) *La giungla dei bilanci familiari* (Bologna: Il Mulino).

Gramaglia, M. (1974) 'Referendum e liberazione della donna', in L. Castellina (ed.), Famiglia e società capitalistica, *Il Manifesto*, 1 (Rome: Alfani Editore).

Guillén, A. M. (1992) 'Social Policy in Spain: From Dictatorship to Democracy (1939–1982)', in Z. Ferge and J. E. Kolber (eds), *Social Policy in a Changing Europe* (Frankfurt-on-Main: Campus Verlag, and Boulder: Westview Press), pp. 119–42.

Guillén, A. M. (1996) 'Welfare State Development in Spain: An Historical and Explanatory Approach', paper presented at the MIRE Conference, Florence, EUI, 21–24 February).

Hantrais, L. (1994) 'Comparing Family Policy in Britain, France and Germany', *Journal of Social Policy*, 23.2, pp. 135–60.

Hantrais, L. and Letablier, M.-T. (1996) 'Familles, travail et politiques familiales en Europe', *Cahiers du Centre d'Etudes de l'Emploi* (Noisy-le-Grand: Presses Universitaires de France).

Heclo, H. (1974) *Modern Social Politics in Britain and Sweden* (New Haven and London: Yale University Press).

Hernández, F. (1990) 'Estructura ocupacional y mercado de trabajo en España', in S. Giner (ed.), *España. Sociedad y política* (Madrid: Espasa-Calpe), pp. 143–67.

Hernes, H. M. (1989) *Welfare State and Women Power*, Essays in State Feminism (Oslo: Norwegian University Press).

Heywood, P. (1995) *The Government and Politics in Spain* (London: Macmillan).

Hobson, B. (1990) 'No Exit No Voice: Women's Economic Dependency and the Welfare State', *Acta Sociologica*, 33, pp. 235–50.

Hoffner, C. (1940) 'Le development de la legislation sur les allocations familiales au cours des annes recentes', *Revue International de Travail*, 41 (April).

Höllinger, F. and Haller, M. (1990) 'Kinship and Social Network in Modern Societies: A Cross-Cultural Comparison Among Seven Nations', *European Sociological Review*, 6.2, pp. 103–24.

Holman, O. (1996) *Integrating Southern Europe* (London: Routledge).

ICSRI (Istituto Centrale Statistica del Regno d'Italia) (1943) *Annali di statistica*, 7th series, vol. 7 (Rome), pp. 12–13.

Iglesias de Ussel, J. (1994a) 'Redes Familiares', in M. Juárez (ed.), *V Informe sociológico sobre la situación social en España. Sociedad para todos en el año 2000* (Madrid: FOESSA), pp. 481–7.

Iglesias de Ussel, J. (1994b) 'Tensiones y ruptura familiar', in M. Juárez (ed.), *V Informe sociológico sobre la situación social en España. Sociedad para todos en el año 2000* (Madrid:

FOESSA).

Iglesias de Ussel, J. and Flaquer, L. (1993) 'Familia y análisis sociológico: el caso de España', *Revista Española de Investigaciones Sociales*, 61, pp. 57–75.

INE (Instituto Nacional de Estadistica) (1977) *Estadística de la Enseñanza en España, curso 1975–76* (Madrid: Ministerio de Economía).

INE (Instituto Nacional de Estadistica) (1982) *Estadistica de la Enseñanza en España, curso 1979–1980* (Madrid: Ministerio de Economia).

INE (Instituto Nacional de Estadistica) (1991) *Encuesta Sociodemográfica. Resultados nacionales*, vol. 1, *hogar y familia* (Madrid: Ministerio de Economía).

INPS (Instituto Nazionale Previdenza Sociale) (1948) *Mezzo secolo di attività assicurativa e assistenziale, 1898–1948* (Rome: INPS).

IM (Instituto de la Mujer) (1990) *I Plan para la igualdad de oportunidades de las mujeres 1988–1990. Evaluacion* (Madrid: Ministerio de Asuntos Sociales).

IM (Instituto de la Mujer) (1993) *II Plan para la igualdad de oportunidades de las mujeres 1993–1995* (Madrid: Ministerio de Asuntos Sociales).

ISSP (International Social Survey Program) (1986) *Social Network and Support System*, Zentralarchiv fuer Empirische Sozialforshung an der Universität zu Köln.

ISTAT (Instituto Nazionale di Statistica) (1973–1983–1989–1992) *Annuario Statistico Italiano*, edizione anni (Rome).

ISTAT (Instituto Nazionale di Statistica) (1983a) 'Contabilità nazionale. Fonti e metodi', *Annali di statistica*, 20th series, vol. 4 (Rome).

ISTAT (Instituto Nazionale di Statistica) (1983b) *Annuario di Statistica Italiano*, various years (Rome).

ISTAT (Instituto Nazionale di Statistica) (1985) *Indagine sulle strutture ed i comportamenti familiari* (Rome).

ISTAT (Instituto Nazionale di Statistica) (1993) *Indagine Multiscopo sulle Famiglie anni 1987–91*, vol. 2 (Rome).

Jenson, J. (1986) 'Gender and Reproduction. Or, Babies and the State', *Studies in Political Economy*, 20, pp. 9–45.

Jenson, J. (1996) 'Au foyer citoyennes! Gender Relations and Changing Childcare Policy in France', paper presented at the 10th Conference of Europeanists (Chicago: 14–16 March).

Jenson, J. and Sineau, M. (eds) (1998) *Qui doit garder le jeune enfant?* (Paris: Droit et Société LGDJ).

Jurado, T. and Naldini, M. (1996) 'Is the South so Different? Italian and Spanish Families in Comparative Perspective', *South European Society and Politics*, 1.3, pp. 42–66.

Kahn, A. J. and Kamerman, S. B. (1990) 'Methodological Cautions for Comparative Research in Social Protection', in 'International Comparisons of Social Security Policies and Systems'. Proceedings of a MIRE and ISSA conference held in June.

Kamerman, S. and Kahn, A. J. (eds) (1978) *Government and Families in Fourteen Countries* (New York: Columbia University Press).

Kaufmann, F.-X. (1996) 'Les Etats providence européens dans leurs rapports avec la famille', in J. Commaille and F. Sigly (eds), *La famille dans la construction de l'Europe politique*, *Le Chaiers du CEVIPOF*, 15.

Kersbergen, K. (1995) *Social Capitalism. A Study of Christian Democracy and the Welfare State* (London and New York: Routledge).

Knijn, T. and Kramer, M. (1997) 'Gender and the Caring Dimension of Welfare States: Toward Inclusive Citizenship', *Social Politics*, 4.3, pp. 328–61.

Korpi, W. (1978) *The Working Class in Welfare Capitalism* (London: Routledge and Kegan Paul).

Koven, S. and Michel, S. (1993) *Mothers of a New World. Maternalist Politics and the Origins of the Welfare State* (New York: Routledge).

Langan, M. and Ostner, I. (1991) 'Gender and Welfare: Towards a Comparative Framework', in Graham Room (ed.), *Towards a European Welfare State?* (Bristol: Saus).

Lasch, C. (1977) *Haven in a Heartless World* (New York: Basic Books).

Laslett, P. (ed.) (1972) *Household and the Family in the Past Time* (Cambridge: Cambridge University Press).

Laslett, P., Oeppen, J. and Smith, E. (1993) 'La parentela estesa verticalmente dell'Italia del XXI seccolo', *Polis*, 7, pp. 121–39.

Lazzerini, G. (ed.) (1990) *Strutture familiari e tendenze* (Torino: Giappichelli Editore).

Leibfried, S. (1992) 'Towards a European Welfare State? On Integrating Poverty Regimes into the European Community', in Z. Ferge and J. E. Kolberg (eds), *Social Policy in a Changing Europe* (Boulder: Westview Press), pp. 345–79.

Leira, A. (1992) *Welfare States and Working Mothers: The Scandinavian Experience* (Cambridge: Cambridge University Press).

Leira; A. (1993) 'The Women-Friendly Welfare State? The Case of Norway and Sweden', in J. Lewis (ed.), *Women and Social Policies in Europe* (London: Edward Elgar).

Leira, A. (1998) 'Caring as a Social Right: Cash for Child Care and Daddy Leave', *Social Politics* (fall), pp. 362–78.

Lewis, J. (1992) 'Gender and the Development of Welfare Regimes', *Journal of European Social Policy*, 2.3, pp. 159–73.

Lewis, J. (ed.) (1993) *Women and Social Policies in Europe. Work, Family and the State* (London: Edward Elgar).

Lewis, J. (1997) 'Gender and Welfare Regimes: Further Thoughts', *Social Politics*, 24.2 (summer), pp. 160–77.

Lewis, J. (ed.) (1998) *Gender, Social Care and Welfare State Restructuring in Europe* (London: Ashgate).

Lewis, J. and Åström, G. (1992) 'Equality, Difference, and State Welfare: Labor Market and Family Policies in Sweden', *Feminist Studies*, 18.1 (spring), pp. 59–87.

Lewis, J. and Ostner, I. (1994) 'Gender and Evolution of European Social Policies', *ZeS-Arbeitpapier*, 4, Centre for Social Policy Research, University of Bremen.

Litwak, E. (1960) 'Geographic Mobility and Extended Family Cohesion', *American Sociological Review*, 25 (June), pp. 385–94.

Livi Bacci, M. (1977) *A History of Italian Fertility During the Last Two Centuries* (Princeton: Princeton University Press).

Longo, M. (1985) 'Assistenza sociale e assistenza familiare', *Giurisprudenza Italiana*, 6.4, pp. 3–6.

López, A. (1988) *La discriminación de la mujer en el mercado de trabajo español* (Madrid: Ministerio de Trabajo y Seguridad Social).

López López, M. T. (1996) *La protección social a la familia en España y en los demás estados miembros de la Unión Europea* (Madrid: Fundación BBV).

MAIC (Ministro Agricoltura Industria e Commercio) (1906) *Statistica industriale. Riassunto delle notizie sulle condizioni industriali del Regno* (Rome).

Mancina, C. (1981) *La famiglia. Teoria storia e funzioni della comunità familiare* (Rome: Editori Riuniti).

Marcet, G. P. (1993) *Entrar, Quedarse, Avanzar* (Madrid: Siglo Veintiuno Editores).

Martin, C. (1996) 'Social Welfare and Protection by the Family in Southern Europe: Are There Any Specificities?', paper presented at the Comparing Social Welfare Systems in Southern Europe conference, Florence, European University Institute, 22–24 February.

MAS (Ministerio de Asuntos Sociales) (1990) 'La situación de la atención a la primera infancia (0–3 años) en España y posibles modelos a impulsar desde las administraciones publicas', unpublished document (Madrid: Dirección General de Protección Jurídica del Menor).

MAS (Ministerio de Asuntos Sociales) (1992) *La atención a la primera infancia en Europa y España* (Madrid: MAS, Subdirección General de Estudios y Documentación).

McIntosh, M. (1978) 'The State and the Oppression of Women', in A. Kuhn and R. M. Wolpe (eds), *Feminism and Materialism* (London: Routledge and Kegan Paul), pp. 254–89.

McLaughlin, E. and Glendinning, C. (1994) 'Paying for Care in Europe: Is There a Feminist Approach?', in L. Hantrais and S. Morgan (eds), *Family Policy and the Welfare of Women*,

Cross-National Research Papers, 3rd series, 3 (European Research Centre, University of Loughborough).

MEC (Ministerio de Educación y Ciencia) (1988) *Estadística de la Enseñanza en España, Niveles no Universitario, curso 1985–86* (Madrid: MEC, Centro de Publicaciones).

MEC (Ministerio de Educación y Ciencia) (1994) *Estadística de la Enseñanza en España, Niveles no Universitario, curso 1990–91* (Madrid: MEC, Centro de Publicaciones).

Meil, G. (1994) 'Evolution de la politique familiale en Espagne. Du salaire familial à la lutte contre la pauvreté', *Population,* 4.5, pp. 959–83.

Meldini, P. (1975) *Sposa e madre esemplare* (Firenze: Guaraldi Editore).

Melograni, P. (ed.) (1988) *La famiglia Italiana dall'Ottocento ad oggi* (Bari: Laterza).

Menniti, A. *et al.* (1992) 'Changes in Family Life in the 1980s. The Italian Case', paper presented at the 2nd working group meeting on Family Policy and Family Life, Bonn, April.

Merli, S. (1972) *Proletariato di fabbrica e capitalismo industriale. Il caso italiano (1880–1900)* (Firenze: La Nuova Italia).

Millar, J. and Warman, A. (1996) 'Defining Family Obligations in Europe. The Family, the State and Social Policy' (Report to the Rowntree Foundation, York).

Millar, J. andWarman, A. (1996b) *Family Obligations in Europe* (London: Family Policy Studies Centre).

Millet, C. J. (1992) 'Fiscalidad y familia. El estado de la cuestión', *Infancia y Sociedad,* 16, pp. 171–8.

Mingione, E. (1994) 'Family Structures and Family Strategies: Confronting a Changing Division of Labour in Different European Contexts', paper presented at the Gender and the Use of Time forum, Badia Fiesolana, European University Institute.

MEH (Ministerio de Economía y Hacienda) (1994) *Estadísticas Presupuestarias y Fiscales* (Madrid: MEH, Secretaria General Técnica).

MISSOC (Mutual Information System on Social Protection) (1996) *Social Protection in the Member States of the European Union* (Brussels, Luxembourg).

MLPS (Ministero del Lavoro e della Previdenza Sociale) (1983) 'Famiglia e reddito. La redistribuzione monetaria del reddito in funzione delle situazioni familiari', *Commissione nazionale per i problemi della famiglia*, Rapporto del II gruppo di lavoro (Rome: Istituto Poligrafico e Zecca dello Stato).

Moreno, L. (1997) *The Spanish Development of Southern Europe*, working paper 97–04 (Madrid: IESA–CSIC).

Moreno, L. and Arriba, A. (1998) 'Decentralization, Mesogovernments, and the New Logic of Welfare Provision in Spain', conference paper WS 17, European University Institute (Florence: 11–12 December).

Moreno, L. and Sarasa S. (1992) *The Spanish 'Via Media' to the Development of the Welfare State*, working paper 92–13 (Madrid: IESA–CSIC).

Moreno, L. and Sarasa S. (1993) 'Genesis y desarrollo del Estado del Bienestar en España', *Revista Internacional de Sociología*, 6, pp. 27–69.

Morisi, M. (1992) *Le leggi del consenso. Partiti e interessi nei primi parlamenti della Repubblica* (Soveria Mannelli: Rubettino).

Moroney, R. M. (1976) *The Family and the State: Considerations for Social Policy* (Harlow: Longmans).

Moss, P. (1990) *La atención a la infancia en las Comunidades Europeas (1985–1990)* (Brussels: Cuadernos de Mujeres de Europa, Comisión Europea).

Mottura, G. and Pugliese, E. (1974) *Agricoltura, mezzogiorno e mercato del lavoro* (Bologna: Società Editrice Il Mulino).

Muro, J., Raymond, J. L., Toharia, L. and Uriel, E. (1988) *Análisis de las condiciones de vida y trabajo en España* (Madrid: Ministerio de Economía y Hacienda, Secretaría de Estado de Economía).

Musatti, T. (1992) *La giornata del mio bambino. Madri, lavoro e cura dei più piccoli nella vita quotidiana* (Bologna: Il Mulino).

MZES (Mannheimer Zentrum für Europaische Sozialforshung) (1994) *OECD Data Base* (Mannheim: August).

Naldini, M. (1999) 'Evolution of Social Policy and the Institutional Definition of Family Models. The Italian and Spanish Cases in Historical and Comparative Perspective', D.Phil. thesis (Florence: European University Institute).

Naldini, M. (2000) 'Family Allowances in Italy and Spain: Long Ways to Reform', in T. Bahle and A. Pfenning (eds), *Family Policy Patterns* (Frankfurt-on-Main: Peter Lang Verlag der Wissenschaften).

Nash, M. (1983) *Mujer, Familia y trabajo en España, 1875–1936* (Barcelona: Anthropos Editorial del Hombre).

Nash, M. (1991) 'Pronatalism and Motherhood in Franco's Spain', in G. Bock and P. Thane (eds), *Maternity and Gender Policies* (London and New York: Routledge).

Norvez, A. (1990) 'De la naissance a l'école. Santé, modes de garde et pré scolarité dans la France contemporaine', *Cahier PUF–INED*, 126.

Nuñez Perez, M. G. (1994) 'La implantación y los resultados del seguro de maternidad en la segunda republica', in Instituto Universitario de Estudios de la Mujer (ed.), *Ordenamiento jurídico y realidad social de la mujeres, siglos XVI a XX,* Actas de las cuartas jornadas de investigación interdisciplinaria (Madrid: University of Madrid), pp. 363–75.

O'Connor, J. S. (1993) 'Gender, Class and Citizenship in the Comparative Analysis of Welfare State Regimes: Theoretical and Methodological Issues', *British Journal of Sociology*, 44.3, pp. 501–18.

O'Connor, J., Orloff, A. and Shaver, S. (1999) *States, Markets and Families, Liberalism and Social Policy in Australia, Canada, Great Britain and the United States* (Cambridge: Cambridge University Press).

OECD (Organisation for Economic Co-operation and Development) (1980, 1989, 1991) *National Account of OECD Countries* (Paris: OECD).

OECD (Organisation for Economic Co-operation and Development) (1991) *The Tax/Benefit Position of Production Workers, 1987–1990* (Paris: OECD).

OECD (Organisation for Economic Co-operation and Development) (1994) *The OECD Jobs Study. Evidence and Explanations.* Part 1, *Labour Market Trends and Underlying Forces of Change* (Paris: OECD).

OECD (Organisation for Economic Co-operation and Development) (1995) *National Accounts. Main Aggregates*, vol. 1, 1960–1993 (Paris: OECD).

Orloff, A. S. (1993) 'Gender and the Social Rights of Citizenship: The Comparative Analysis of Gender Relations and Welfare State', *American Sociological Review*, 58, pp. 303–28.

Orloff, A. S. (1996) 'Gender and the Welfare State', *Working Paper 1996–79* (Madrid: Instituto Juan March de Estudios e Investigaciones).

Paci, M. (1980) *Famiglia e mercato del lavoro in una economia periferica* (Milan: Angeli).

Paci, M. (1981) *Mutamento e classi sociali in Italia* (Napoli: Liguori Editore).

Paci, M. (1984a) 'Il sistema italiano di welfare tra tradizione clientelare e prospettive di riforma', in U. Ascoli (ed.), *Welfare state all'Italiana* (Bari: Laterza), pp. 297–326.

Paci, M. (1984b) *La struttura sociale Italiana* (Bologna: Il Mulino).

Paci, M. (1987), 'Long Waves in the Development of Welfare Systems', in S. C. Maier (ed.), *Changing Boundaries of the Political* (Cambridge: Cambridge University Press).

Paci, M. (1989) *Publico e privato nei moderni sistemi di welfare* (Napoli: Liguori Editore).

Palacios, J. (1989) 'Contextos de crianza y educación de los niños españoles menores de seis años', *Infancia y Aprendizaje*, 46, pp. 83–116.

Pasquino, G. (1996) 1946–1995. *Profilo della politica in Italia* (Roma-Bari: Editori Laterza).

PCI (Partito Comunista Italiano) (1976) 'Il Partito Comunista Italiano', in *Le promesse dei partiti, Biblioteca della libertà*, vol. 13, pp.160–98.

PCI (Partito Comunista Italiano) (1979) 'Il programma dei comunisti per l'VIII legislatura', *Unità*, special issue, 8.

PDS (Partito Democratico di Sinistra) (1992) *Costruiamo una nuova Italia. PDS opposizione che costruisce* (party programme).

PDS (Partito Democratico di Sinistra) (1994) *Programma di governo del PDS. Per ricostruire un'Italia più giusta, più unita, più moderna* (party programme).

Pedersen, S. (1993) *Family, Dependence, and the Origins of the Welfare State. Britain and France 1914–1945* (Cambridge: Cambridge University Press).

Pérez Yruela, M. (1990) 'La sociedad rural', in S. Giner (ed.), *España. Sociedad y política* (Madrid: Espasa-Calpe), pp. 199–243.

Pérez Yruela, M. and Giner, S. (eds) (1988) *El Corporatismo en España* (Barcelona: Ariel Sociología).

Piccone Stella, S. (1997) 'I giovani in famiglia', in M. Barbagli and C. Saraceno (eds), *Lo stato delle famiglie in Italia* (Bologna: Il Mulino), pp. 151–62.

Picontó-Novales, T. (1996) 'Family Law and Family Policy in Spain', in J. Kurczewski and M. Maclean (eds), *Family Law and Family Policy in the New Europe* (Dartmouth: Oñati International Institute for the Sociology of Law), pp. 109–27.

Pierson, P. (1993) 'The Creeping Nationalization of Income Transfers in the United States, 1935–94', in S. Liebfried and P. Pierson (eds), *European Social Policy* (Washington, DC: Brookings), pp. 301–28.

Pierson, P. (1994) *Dismantling the Welfare State? Regan, Thatcher, and the Politics of Retrenchment* (Cambridge: Cambridge University Press).

Pitch, T. (1983) 'Ruolo delle donne e mutamento delle strategie di controllo sociale', *Dei delitti e delle pene*, 1, pp. 94–109.

Pocar, V. and Ronfani, P. (1995) 'Italie: la stratégie des convergences', in L. Assier-Andrieu and J. Commaille (eds), *Politiques des Lois en Europe. La filiation comme modèle de comparaiso (*Paris: LGDJ), pp. 185–226.

Pocar, V. and Ronfani, P. (1998) *La famiglia e il diritto* (Roma-Bari: Editori Laterza).

PSI (Partito Socialista Italiano) (1968) 'Il programma socialista', *Avanti*, 28 (April).

PSOE (Partido Socialista Obrero Espanañol) (1982) *Por el cambio, Programa Electoral Partido Socialista Obrero* (party programme).

PSOE (Partido Socialista Obrero Espanañol) (1989) *España en progreso. Programa Electoral 1989* (party programme).

Rapoport, R. (1989) 'Ideologies About Family Forms: Towards Diversity', in K. Boh (ed.), *Changing Patterns in European Family Life* (London and New York: Routledge).

Rimlinger, G. (1971) *Welfare Policy and Industrialization in Europe, America and Russia* (New York: John Wiley).

Rodríguez Cabrero, G. (1989) 'Orígines y evolución del Estado de Bienstar español en su perspectiva histórica. Una visión general', *Politica y Sociedad*, 2, pp. 79–87.

Rodríguez Cabrero, G. (1990) *El gasto público en servicios sociales en España (1972–1988)* (Madrid: Ministerio de Asuntos Sociales).

Rodríguez Cabrero, G. (1992) 'Welfare Policies in Spain: Main Problems Under the Impact of the European Integration', working paper prepared for the research conference From European Societies to European Society: the National Welfare States and European Integration (Saint Martin: November).

Rodríguez Cabrero, G. (1994) 'La política social en España: 1980–92', in *V Informe Sociológico sobre la Situación Social en España* (Madrid: Fundación FOESSA), pp. 1443–94.

Roussel, L. (1983) 'Les ménages d'une personne: l'évolution récente', *Population*, 38.6, pp. 995–1016.

Roussel, L. (1992) 'La famille en Europe occidentale: divergences et convergences', *Population*, 1, pp. 133–52.

Roussel, L. (1994) 'La famille en Europe occidentale depuis 1950', paper presented at the international conference, Changes in Family Patterns in Western Countries (Bologna: Comune di Bologna Assessorato alle politiche sociali e dell'immigrazione, 6–8 October).

Rowland, R. (1987) 'Matrimonio y familia en el mediterráneo occidental: algunas interrogaciones', in F. Chacón (ed.), *Familia y sociedad en el mediterráneo occidental. Siglos XV–XIX* (Universidad de Murcia).

Ruggie (1984) *The State and the Working Mothers* (Princeton: Princeton University Press).

Rull Sabater, A. (1971) *La seguridad social en España*, vol. 1 (Madrid: Euramerica).

Sabbadini, L. (1994) 'Reti di relazioni familiari e tipologia della famiglia', in Comitato per l'anno Internazionale della Famiglia, *Per una Politica Familiare in Italia* (Rome: Presidenza del Consiglio dei Ministri, Ministero per gli Affari Sociali), pp. 173–89.

Sabbadini, L. (1999) 'Modelli di formazione e organizzazione della famiglia', paper presented at Le Famiglie Interrogano le Politiche Sociali, Bologna, 29–31 March (published by Presidenza del Consiglio dei Ministri.

Sabbadini, S. (1991) 'Le unioni libere', in A. Menniti (ed.), *Le famiglie italiane degli anni 80*, Monographs IRP/2 (Rome).

Sainsbury, D. (ed.) (1994a) *Gendering Welfare States* (London: Sage).

Sainsbury, D. (1994b) 'Women's and Men's Social Rights: Gendering Dimensions of Welfare States', in D. Sainsbury (ed.), *Gendering Welfare States* (London: Sage), pp. 150–68.

Sainsbury, D. (1996) *Gender, Equality and Welfare States* (Cambridge: Cambridge University Press).

Salas, R. and Perez Villacosta, E. (1992) 'La evolucion de los cambios legales de la tarifa e deduciones del Irpef y sus repercusiones sobre el tipo efectivo 1982–1989', *Papeles de Trabajo*, 19 (Madrid).

Salvadori, M. (1982) *Storia dell'Età contemporanea. Dalla restaurazione all'Eurocominismo*, vol. 2, 1914–1945 (Torino: Loescher Editore).

Samaniego Boneu, M. (1988) *Los seguros sociales en la España del siglo XX. La unificación de los seguros sociales. A debate. La segunda República* (Madrid: Ministerio de Trabajo y Seguridad Social).

Sapelli, G. (1995) *Southern Europe Since 1945* (Harlow: Longman).

Saraceno, C. (1984) 'Il rapporto famiglia-Stato e i contributi dalla parte delle donne', *Inchiesta*, 65, pp. 26–32.

Saraceno, C. (1991) 'Redefining Maternity and Paternity: Gender, Pronatalism and Social Policies in Fascist Italy', in G. Bock and P. Thane (eds), *Maternity and Gender Policies* (London and New York: Routledge).

Saraceno, C. (1994) 'The Ambivalent Familism of the Italian Welfare State', *Social Politics*, 1, pp. 60–82.

Saraceno, C. (1996) *Sociologia della famiglia* (Bologna: Il Mulino).

Saraceno, C. (1998) *Mutamenti della famiglia e politiche sociali in Italia* (Bologna: Il Mulino).

Saraceno, C. (2000) 'Politiche sociali e famiglie', in M. Barbagli and D. Kertzer (eds), *The History of European families* (New Haven and Laterza: Yale University Press).

Sarasa, S. and Moreno, L. (1995) *El estado del bienestar en la Europa del sur* (Madrid: Consejo Superior de Investigaciones Científicas).

Schultheis, F. (1992) 'L'avenir de la famille au centre des antinomies de la modernité', in Haut Conseil de la population et de la famille, *Du politique et du social dans l'avenir de la famille* (Paris: Documentation Française).

Sgritta, G. B. (1986) 'La struttura delle relazioni inter-familiari', in ISTAT, *Annali di Statistica*, 9.6).

Sgritta, G. B. (1989) 'Towards a New Paradigm: Family in the Welfare State Crisis', in K. Boh (ed.), *Changing Patterns in European Family Life* (London and New York: Routledge).

Siim, B. (1987) 'The Scandinavian Welfare states. Towards Sexual Equality or a New Kind of Male Domination?', *Acta Sociologica*, 30.3–4, pp. 255–70.

Skocpol, T. (1985) 'Bringing the State Back In', in P. B. Evans, D. Rueschemeier and T. Skocpol (eds), *Bringing the State Back In* (Cambridge: Cambridge University), pp. 3–43.

Skocpol, T. and Amenta, E. (1986) 'States and Social Policies', *Annual Review of Sociology*, 12, pp. 131–57.

Skocpol, T. (1992) *Protecting Soldiers and Mothers* (Cambridge, MA: Harvard University Press).

Solsona, M. (1991) 'The Problem of Measuring Women's Work in Spain', *Iberian Studies*, 20.1–2, pp. 6–28.

Soto Carmona, A. (1989) *El trabajo industrial en la España contemporánea. 1874–1936* (Barcelona: Anthropos Editorial del Hombre).

Spicker, P. (1991) 'The Principle of Subsidiarity and the Social Policy of the European Community', *Journal of European Social Policy*, 1.1, pp. 3–14.

Sterdyniak, H. (1994) 'In Defence of the Quotient Familial' in Mire (Mission Research) (ed.), Comparing the Social Welfare System in Europe, Oxford Conference, vol. 1, pp. 369–410.

Svimez (1961) *Un secolo di Statistiche Italiane Nord e Sud (1861–1961)* (Rome).

Titmuss, R. (1974) *Social Policy* (London: Allen and Unwin).

Togliatti, P. (1965) *L'emancipazione femminile: discorsi alle donne* (Rome: Editori Riuniti).

Tomás, F. and Valiente, G. (1992) *Manual de historia del derecho español* (Madrid: S. A. Teconos).

Torres, J. A. (1993) 'Análisis empírico de las desigualdades educativas en España', in *I Simposio sobre la igualdad y distribución de la renta y la riqueza* (Madrid: Fundación Argentaria).

Trifiletti, R. (1984) 'La famiglia italiana e i servizi sociali. Alcune inconciliabilità negli schemi teorici emergenti', *Inchiesta*, 65 (Aug.–Sept.), pp. 11–24.

Trifiletti, R. (1995a) 'Family Obligation in Italy', *Bath Social Policy Papers*, 23 (Bath: University of Bath).

Trifiletti, R. (1995b) 'The Gendered "Rationalization" of Italian Social Policies in the Nineties', paper presented at the 2nd ESA conference, European Societies: Fusion or Fission (Budapest: 30 Aug.–2 Sept.).

UCD (Unión de Centro Democrático) (1977) *Manual para veintidos millones de electores* (party programme).

UCD (Unión de Centro Democrático) (1979) *Programa electoral de UCD* (party programme).

Valero, A. (1992) 'La prevalecía de la familia nuclear en el sistema familiar español', *Revista Internacional de Sociología. Tercera época*, 3 (Sept.–Dec.), pp. 183–210.

Valero, A. and Lence, C. (1995) 'Nupcialidad, fecundidad y familia. La paradoja del comportamiento de la nupcialidad y la fecundidad en España', *Revista Internacional de Sociología*, Tercera Época, 11, pp. 89–114.

Valiente, C. (1995) 'Children First: Central Government Child Care Policies in Post-Authoritarian Spain (1975–1994)', in J. Brannen and M. O'Brien (eds), *Childhood and Parenthood* (London: Institute of Education, University of London), pp. 249–66.

Valiente, C. (1996a) 'The Rejection of Authoritarian Policy Legacies: Family Policy in Spain 1975–1995', *South European Society and Politics*, 1.1 (summer), pp. 95–114.

Valiente, C. (1996b) 'Women in Segmented Labour Market and Continental Welfare States: The Case of Spain', in L. Hantrais and M. T. Letablier (eds), *Comparing Families and Family Policies in Europe*, Cross-National Research Papers, 4th series (Loughborough: Cross-National Research Group).

Van Kersbergen, K. (1995) *Social Capitalism. A Study of Christian Democracy and the Welfare State* (London and New York: Routledge).

Villac, M. (1993) 'La politique familiale, débats autour d'une définition', in Politique familiale et transformation des modes de vie, *Solidarité Santé-Etudes statistiques,* 4 (Oct.–Dec.).

Vincenzi Amato, D. (1973) *Gli Alimenti* (Milan: Giuffré Editore).

Vincenzi Amato, D. (1988) 'La famiglia e il diritto', in P. Melograni (ed.), *La famiglia italiana dell'Ottocento ad oggi* (Roma-Bari: Laterza), pp. 629–700.

Wallerstein, I. (1979) *The Capitalist World Economy* (Cambridge: Cambridge University Press).

Weir, M., Orloff, A. and Skocpol, T. (eds) (1988) *The Politics of Social Policy in the United States* (Princeton: Princeton University Press).

Wennemo, I. (1992) 'The Development of Family Policy. A Comparison of Family Benefits and Tax Reduction for Families in 18 OECD Countries', *Acta Sociologica*, 35.3, pp. 201–17.

Wilensky, H. L. (1975) *The Welfare State and Equality. Structural and Ideological Roots of Public Expenditures* (Berkeley: University of California Press).

Wilensky, H. L., Luebbert, G. M., Hahn, S. R. and A. M. Jamienson (1985) *Comparing Social Policy: Theories, Methods, Findings* (Berkeley: University of California Press).

Zamagni, V. (1994) 'A Century of Change: Trends in the Composition of the Italian Labour Force, 1881–1981', in G. Federico (ed.), *The Economic Development of Italy Since 1870* (Cambridge: University Press Cambridge), pp. 210–71.

INDEX

Please note that references to footnotes are denoted by the letter 'n' and number of note appearing after the page number. References to figure or tables are in italic print.

I.D. No. 1410495
UNIVERSITY OF BRADFORD
LIBRARY
- 2 JUL 2004
ACCESSION No. 0344236640
CLASS No.
LOCATION

UNIVERSITY OF BRADFORD
WITHDRAWN
AND
DISCARDED
LIBRARY